Bernina Zehioua Hecham

# THE URBAN PROJECT: MEANINGS, STRATEGIES, DERIVATIVES AND EXPERIENCES

Bernina Zehioua Hecham

# THE URBAN PROJECT: MEANINGS, STRATEGIES, DERIVATIVES AND EXPERIENCES

## Urban project approach

ScienciaScripts

**Imprint**



Cover image: www.ingimage.com

This book is a translation from the original published under ISBN 978-620-6-71513-9.

Publisher:
Sciencia Scripts
is a trademark of
Dodo Books Indian Ocean Ltd. and OmniScriptum S.R.L publishing group

120 High Road, East Finchley, London, N2 9ED, United Kingdom
Str. Armeneasca 28/1, office 1, Chisinau MD-2012, Republic of Moldova, Europe
Managing Directors: Ieva Konstantinova, Victoria Ursu
info@omniscriptum.com

Printed at: see last page
**ISBN: 978-620-8-55849-9**

## Contents

## FOREWORD

This collection is a course aid for the Urban Project **subject**, intended for students in the Master 1 Urban Planning programme, Faculty of Architecture and Urban Planning, CONSTANTINE 3 University, Algeria. It constitutes a course manual and subjects dealt with in TD on the "urban project" approach. Nevertheless, because of the importance of its content, it covers the strategies used in the "URBAN PROJECT" approach and provides theoretical support for any student or reader wishing to learn more about the URBAN PROJECT CONCEPT.

This compendium is structured in TWO SECTIONS: an acquired knowledge section containing seven courses and a tutorial section divided into two sub-sections: the first concerns tutorial work in the form of lectures on urban planning operations around the world and the second sub-section concerns tutorial work in the form of lectures on urban planning in Algeria. We wanted to highlight our students' home towns, where they can rub shoulders with the local authorities and citizens of their town. The study area is his favourite terrain, and he knows it very well given that he it practice on a daily basis.

**A/ ACQUIRED KNOWLEDGE SECTION :**

1. LESSON 1/ **looking at the foundations of the urban project**
2. COURSE No. II/ **looks at the scales affected by the approach, depending on the context and the importance of the actions**
3. COURSE No. III **focuses on the process and the interdisciplinary approach involved**
4. LESSON N°IV/ **try to give some definitions because of the ambiguity of the concept /complementarity**
5. COURSE N°V/ **looks at the players involved and the urban planning actions targeted.**
6. COURSE N°VI/ **looks at concurrent or transactional town planning as a middle way.**
7. COURS N°VII/ **is interested in the sustainable city and participative urban planning, the spearhead of the urban project.**

**B/ TUTORIAL SECTION**

It is divided into two sub-sections: firstly, international subjects on operations carried out in cities around the world in the name of the urban project studied by students of the master 1 urban planning and the second sub-section focuses on the urban project in Algeria, where urban planning actions in different Algerian cities are highlighted. These students will each have the opportunity to talk to the local authorities of their town (survey method, interview technique or interview with the decision-makers) and to talk to the residents (survey method, observation technique and questionnaire) to find out what they think of these projects and whether they have been consulted.

## SECTION A. KNOWLEDGE ACQUIRED

**1. <u>COURSE NO. 1: FOUNDATIONS OF THE URBAN PROJECT</u>**

**INTRODUCTION**

**2. THE FOUNDATION**

**2.1. BACKGROUND TO PLANNING AND HOW IT HAS EVOLVED AROUND THE WORLD...**

**2.2. RECLAIMING DERELICT LAND AND URBAN REGENERATION**

**CONCLUSION**

**INTRODUCTION**

**THE URBAN PROJECT is based firstly on the PROJECT concept and secondly on the URBAN concept. The interaction between these two concepts, which has an impact on a third element, the stakeholder in all its forms, gives rise to different approaches depending on the discipline and the profile of the stakeholder.**

When DEDALE, builder of the labyrinth and mythical father of architects, passed on to his son ICARUS the power to transgress, through "invention", the laws laid down by man and to free himself from the constraints imposed by NATURE, he took care to give him two recommendations which set the "mortal" limits of his enterprise:

- Don't fly too high, or the sun will melt the wax on your wings;
- Do not fly too low near the sea, as the water will weigh down its feathers.

The mythological account adds that ICARE, intoxicated by his new power, forgot his father's warnings. He could not restrain himself from flying too high and lost his life as a result (**CABET E.,** 1840).

The same applies to the urban project strategies.

***Projectum: Latin .... To throw forward. A project is a set of ideas, intentions and coordinated actions that we decide to carry out over time to achieve an objective. This "projection" calls on human, material and financial resources.***

***(logistical) and budgetary aspects. It is based on an approach, method or process.***

***Urbain: Latin ...Urbs...VILLE, Polis, Cite.toute intervention touchant la ville et sa structure rentre dans le domaine de l'urbain, de la ville*** (**BOUCHAREB**, 2012, P.4.)

This course highlights the basis of the urban project, recalls the circumstances of the birth of this concept, attempts to embed the issues to which this "suitcase" word alludes and sets out the global context of the new urban planning.

**I.THE FOUNDATION**

THE URBAN PROJECT is a polysemic concept that has not yet been well defined and is wide open.

Cities have always been built around specific projects. These bear witness to the "urban creation of an era, the expression of its problems, its management methods and its solutions" (**ARNAUD,**1996, p.3).

In the current urban context, the projects that are transforming the city are characterised by their scale and complexity. The number of sites involved is significant, as is the spatial and financial scale of the projects. These projects often have considerable implications on different scales - political, physical, economic, social and spatial.

*"The term urban project implies a very specific concept, an alternative to the traditional planning in a socio-economic context that has been transformed. The emergence of this concept is part of a process of reversal of ideas which, over the last three decades, has*

*completely renewed the concepts used in urban planning* **(THOMAS**, 1998, p.17).
For his part, **INGALLINA** (2001, p.3) points out that thinking about the urban project is still in vogue and continues to develop. "*It is a concept and a way of acting in the making which mark a moment of transition between the traditional way of thinking about urban planning and a new approach, less fixed and more open to change and debate".*

## 1.1. BACKGROUND TO PLANNING AND HOW IT HAS EVOLVED AROUND THE WORLD.

The term 'urban project' was first used in the 1970s **(THOMAS**, 1998, **INGALLINA**, 2001), in opposition to functionalist urban planning in a changing socio-economic context. In order to situate the emergence of the 'urban project' concept and to understand the issues to which it is attempting to respond, a reminder of the current context which characterises planning is in order.
This new way of intervening in the city is not a fad that will fade away later; on the contrary, it has different architectural, urban planning, political, institutional, economic, financial, spatial and social dimensions....It affects all the foundations and the very basis of traditional planning.

## 1.2. REGENERATION OF DERELICT LAND OR URBAN REGENERATION.

"This is the new way of intervening in the city. It is no longer a question of regulating, guiding and controlling growth on the outskirts of the city, as was done during the Second World War and the post-independence period in Algeria, but of finding initiatives to stimulate the development of central areas and slow down the spread of the urban sprawl. This is what **CHALINE** (1999) calls "regeneration urbanism", which is characterised by the spatial extent of the formation of wasteland, the obligation which forces the public authorities to intervene to reclaim these areas, and the renewal of planning objectives and methods **(COURCIER**, 2002).
Recycling urban spaces is nothing new, but the slowdown in demographic and economic growth in Europe, the property crisis of the 1990s, and the disappearance of certain sectors of activity, leading to businesses closing down or being relocated, have led to a widespread phenomenon of brownfield sites. These are large areas in central districts or close to the historic centre. If these areas represent land and opportunities, the chance to rethink the city is there for the taking.
The concept of the urban project originated in Europe, specifically in Italy, with the restructuring of the city of Bologna. The "recuppero" operation, which began in the 60s, aimed to rebuild the links between the different parts of the city of Bologna, seeking to maintain a balance between them. The city was seen as a whole, and change in one part had immediate and long-term consequences for other parts of the city. The theoretical foundation on which this notion was based was enshrined in the thinking of GUSTAVO GIOVANNONI, who advocated not just giving importance to a historic monument or an old town centre and neglecting the overall context in which that element was located, but taking the city as a whole, as a heritage that needed to be developed. The approach then becomes one of global reflection on the city, leading to an approach that is open to debate and consultation with all the players who make up the city.
In France and in Europe at that time, there was traditional planning. Italy had always been a pioneer in architecture and art. The way in which cities were planned became obsolete because it did not take into account what already existed, and people began to criticise the "clean slate" and renovation. Town planning was criticised for erasing history to the detriment

of the new, and the opinions of citizens, associations and historians were taken into account... Town planning, which used to be the prerogative of the State, became a matter for the tones... Concertation was taken into account, as were the interactions between the various players in the city. Residents and users, who until then had been subjected to state solutions, became responsible players, contributing their ideas to the search for the ideal solution.
This operation in Bologna spread and the HALLES in Paris took advantage of it for their transformation. There were ready-made solutions, but this state of affairs was avoided, and the operation was started gradually "as it went along", out of the criticisms and cultural debates of the Parisians. For the first time, the "genie des lieux" was mentioned.
However, these operations are complex and a burden for towns and cities because of the various consequences they entail. There are many obstacles to overcome: contaminated soil, deterioration of buildings, loss of fiscal potential for local authorities, high redevelopment costs. These abandoned areas have a history, a meaning, they are part of a district, part of a town. And they can no longer be redeveloped on the basis of a clean slate. A number of values have become essential, such as :
-environment, landscape, quality of life, heritage, sustainable development, etc. and the system of players involved is broader and more complex.
Secondly, the problems faced by these sites occur at different scales, from the neighbourhood to the city, the region, the country and even the planet.
The authorities that were supposed to intervene in these areas do not necessarily have the financial capacity to do so. In fact, *"today, the regulatory mechanisms that ensured continuity are proving insufficient to reallocate, in the long term, the voids, gaps and wastelands that open up and blight many urban fabrics. This is the result of an accentuated phase shift between the rhythms and times of the city, those of its built environment and those of its various users"* (**CHALINE**, 1999, p.3).
So, faced with this inability to act with the old means, the authorities have to come up with solutions to attract investment and job-creating businesses, by finding a position in a context of increased competition between cities.
The logic between demand and supply of space has been reversed. In the past, there was a need (for equipment, housing, etc.) and a place had to be found to meet it. Today, a space is available and we need to find a new use for it.
Setting up these brownfield redevelopment projects is a long and difficult process. Developers can no longer envisage designing a plan and implementing it strictly as it has been drawn up. It has to be adapted to a rapidly changing environment and to a complex system of players, which has given rise to new challenges for planners at different scales, from the neighbourhood to the city, to the regional, national and even international level. **THE URBAN PROJECT ATTEMPTS TO RESPOND TO THESE NEW CHALLENGES...**
The urban project is certainly a late concept, but it has become essential in the context of reclaiming spaces that have been overtaken by the circumstances of the century. It calls for the participation of everyone (technocrats, politicians, technicians, other players from different disciplines, users, residents, citizens, etc.), because everyone is concerned by the space and its project. The image created by the project in these areas has an impact on the whole city and its visitors. This overall vision is of interest to the future because it will be bequeathed as a heritage to other generations and the work will be judged, commented on, appreciated or depreciated depending on the impact it has had. Responsibility is shared

**CONCLUSION**

The first course on the urban project enables the learner to retrace the birth of the concept and the operations which marked its birth. A few reminders and preliminaries were therefore necessary to introduce the concept in a flexible way (qualities of the urban project) to the vision of traditional planning and the search for conciliatory and non-segregative strategies for all users of space, whether they be residents, decision-makers, developers, financiers or visitors. As the urban project is polysemic, we will first describe the field of action in which it evolves. We have seen that the urban project is an approach that covers all scales: from the building, through the district, to the whole city, and ultimately encompasses the whole territory. This is the subject of the next course, which looks at the urban scales crossed by the urban project.

**11. COURSE N°II / URBAN PROJECT SCALES**

**INTRODUCTION**

**12. THE SCALES OF THE URBAN PROJECT// THE RELEVANT PERIMETER**

**12.1. THE URBAN PROJECT ON THE SCALE OF THE CONURBATIONATTRACTIVENESS STRATEGY**

**12.2. CITY PROJECTS //IMPACT ON IMAGE**

**12.3. THE OPERATIONAL PROJECT OR THE ARCHITECTURAL PROJECT/ REDEFINING AND COMPOSING/FLEXIBILITY**

**12.4. THE TIME SCALE OF THE URBAN PROJECT// TIME AND SPACE TALENT**

**CONCLUSION**

**INTRODUCTION**

Towns expand, grow and consume land according to economic, cultural, social and functional rationales. The different periods of the city's spatial development bear witness to the growth and land-hungry nature of human and industrial settlements. Already cramped within its own walls, the city gradually nibbled away at all the surrounding areas, made up of communes, villages and other conurbations. With the help of transport, a conurbation took shape. Housing estates were built, followed by roads, schools, shopping centres and other facilities to bring these new districts to life, and because it was cheaper for investors to set up on the outskirts or in the suburbs (town centres had to be preserved, the area was protected, there was no free land available, etc.). Sprawl proliferates, swallowing up the good farmland that feeds the towns. From this point of view, **"building the city on the city"** or regeneration urbanism takes over and then becomes **"the job to be done, to be recommended",** especially if the state of the site is insalubrious and degraded or abandoned.

This new way of thinking puts the brakes on urban expansion, enhances the value of the existing centre, and helps by reclaiming the large areas of land known as "urban wasteland" left behind by the departure of businesses, the relocation of factories, and the abandonment of buildings that have changed function... If these spaces represent opportunities, the chance to rethink the city is there for the taking. The only problem is that taking action is difficult, with sometimes regrettable consequences. So the practice of urban acupuncture is spreading, focusing on places capable of generating induced effects that avoid erasing the memory of places and the symbolic value with which they have been charged over time.

The urban project crosses the neighbourhood, affects the whole city, impacts the territory and takes its orders from the highest levels of authority. With its dual name of project and urban, it reflects the different timescales and scales of intervention.

## II. THE SCALES OF THE URBAN PROJECT// THE RELEVANT PERIMETER

Identifying the scales of the urban project means knowing the relevance of the perimeter within which it will be carried out, which is the fundamental element in assessing the viability of the project. This means defining in advance the area in which the city will be able to exercise its powers effectively and enable economically viable management. The area chosen must be large enough to ensure that the public or private investment sunk into the project is profitable. It must bring together areas of economic attractiveness, business parks, housing areas, agriculture, industry and commerce, and encompass areas of future development. The relevant perimeter is in fact the area of influence of one or more associated economic poles with a sufficiently diversified offer for the town to be self-sufficient.

From a single building to be constructed, rebuilt or renovated, to a district to be remodelled, to a conurbation that needs to be rethought, the urban project crosses all scales, forming networks on the scale of an area, a region or even a country. It calls into question traditional planning, which needed a **master development perimeter** or neighbourhood perimeter in order to plan. In the words of **DEVILLIERS** (1994, p. 12), "*the urban project is not a profession, but a skill; it is not a procedure, but a process. It is about putting together a multidisciplinary team and adopting an attitude that is open to debate, to new ideas, to constructive criticism... In short, it's about working with what already exists.*

So, in addition to temporal and spatial scales, there is transdisciplinarity and consensus between the different players brought together by the interventions.

Depending on the scale, the urban project takes on different names. **MERLIN AND CHOAY** (1996) distinguish three types of meaning according to the territorial scale: **the political urban project, the operational urban project and the architectural urban project.** **INGALLINA** (2001) classifies projects from the point of view of strategic planning in a bottom-up process: **from the local to the global** and in a top-down consultation process "**from the global to the local".** The whole is interwoven through the articulation and interlocking of activities and scales. FIGURE 1.

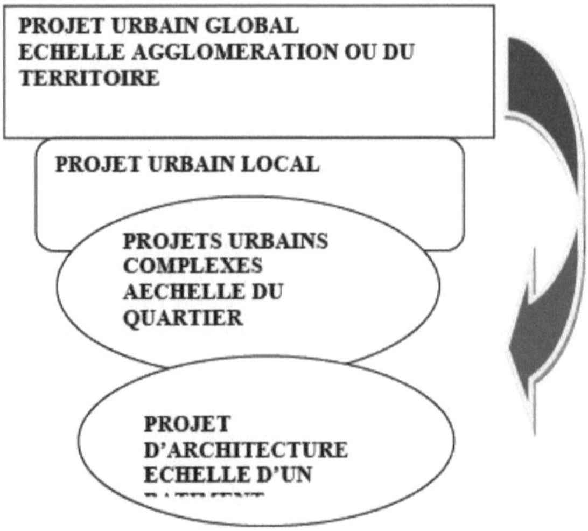

**OVERALL URBAN PROJECT ON A CONURBATION OR REGIONAL SCALE**
**LOCAL URBAN PROJECT**
**COMPLEX URBAN PROJECTS AT NEIGHBOURHOOD LEVEL**
**PROJECT**
**ARCHITECTURE**
**SCALE OF A**

**FIGURE 1: The interweaving of urban project scales (Author, 2023)**

## 11.1. THE URBAN PROJECT FOR THE CONURBATION *//ATTRACTIVENESS* STRATEGY.

The main challenge for towns and cities is to develop in a coherent and balanced way across all their components, whether economic, social, cultural, environmental or land-related.
Traditional development plans were incoherent and did not take account of inter-territoriality. With the urban project, cooperation between neighbouring towns and the same region or municipality comes to the fore, so that by pooling their potential, they can have a major impact on other conurbations. It was therefore necessary to institutionalise cooperation in the field of strategic planning, which **INGALLINA** (2001, p.32) calls *"strategic planning extended to the scale of the conurbation... Strategic planning is the bearer of globality, and its interest lies in attempting to overcome the fragmentation of local issues underlying spatial planning".*
The global approach makes it possible to avoid conflicts between municipalities in the same area, because a preliminary diagnosis is carried out to identify the handicaps and strengths of each municipality, and to identify development opportunities on which strategic planning can be based, targeting the axes likely to drive the area's development forward and taking account of permanent constants and objectives such as sustainable development. A project of limited size (on a neighbourhood scale, for example) can have an impact on the whole town and even the region.

## 11.2. CITY PROJECTS //IMPACT ON IMAGE

The new requirements of the current context, i.e. the recognition of the concept of sustainable development, the affirmation of the project approach in legislative texts and the new organisation of urban areas throughout the world, have given rise to a rigorous methodological approach to the development of city or regional projects. The GPVs (major city projects) are inspired by the reference framework of sustainable development objectives in the territorial diagnosis phase and in the phase of defining the content of the project in order to :

- assessing the region in the light of sustainable development objectives ;
- prioritise a project approach based on local characteristics and issues ;
- to mobilise stakeholders and stimulate debate on the development of the conurbation and a forward-looking vision of its future.

For GPVs to remain relevant today, they need to be updated in line with changes in time and space. They may be subject to cyclical, structural or sectoral changes that mark the evolution of towns and cities, and respond to criteria based on new elements such as :

- the legislative overhaul of territorial planning, which not only offers new tools but is also in line with new political guidelines driven by the requirements of sustainable development;
- the desire expressed by the legislator to give priority to project-based approaches rather than procedures, based on the specific characteristics and challenges of the city in question,

mobilising partners from the design stage through to implementation of the project;

- the imperative of inter-municipal solidarity is gaining ground with the creation of future conurbation contracts.

Given the context, this exercise required a new approach. It was necessary to draw up an inventory of the territory of the conurbation, an inhabited space endowed with an identity, and to highlight its strengths and weaknesses in terms of the challenges of sustainable development. The aim is to ensure that projects are more relevant and that local players take greater ownership of them, by taking a more qualitative and participatory approach.

### 11.3. THE OPERATIONAL PROJECT OR THE ARCHITECTURAL PROJECT REDEFINING AND COMPOSING/FLEXIBILITY

The purpose of an urban project on the local scale of a district or a building is to requalify, recompose, divide up, open up, revitalise, regenerate or sew together an area to be stitched together, a place, a facility, etc. Combining rehabilitation and demolition, renovation and conservation, it must respond to unforeseen circumstances in a spirit of flexibility and adaptation, while mobilising the population around the projects to recreate citizenship and ensure consultation. It draws up the programme on the basis of negotiations with all the stakeholders, working back and forth to ensure coherence and a satisfactory representation of the whole. Working closely with the elements of the surrounding urban form, it ensures that the new project is integrated into its surroundings without detracting from the identity that has already been acquired, or reinforcing and enhancing this aspect. In this sense, as emphasised by **MANGIN AND PANNERAI** (1999), it goes beyond the strict domain of building technicians to encompass the social, economic and cultural aspects of the site. **HUET** (1986) refers to the urban project as an instrument of mediation between the city and architecture.

### 11.4. THE TIME SCALE OF THE URBAN PROJECT// TIME AND TALENT

The urban project is a process which requires the contribution of several disciplines, of several players directly or indirectly involved, taking into account the different parameters of the site, or those of the economy or those relating to the timescale. For the urban project to be realistic, the successive phases of planning and the short, medium and long-term outlook must be taken into account. These phases determine the objectives to be achieved and, for others, the choice of broad, consensual and time-bound guidelines. This factor, "time", if it is well mastered from the outset, encourages investors and especially private partners to invest, because they assess their gain on a time scale of **"return on investment in the short, medium and long term", profit and loss.**

To escape the vacuity of words, the urban project is assessed in time and space, crosses scales flexibly, responds to the narrow-mindedness of traditional planning, sets itself precise, clear, concise objectives, takes on the qualities of flexibility and adaptability, is built with serious and diversified partners, motivated and aware players (public, private, citizens), takes a range of spatial and temporal boundaries, changes obsolete tools by creating others that can cope with the new demands of the processes of 'making the city'.

### CONCLUSION

As the urban project is still a vague concept and its definition merits transdisciplinary exploration, we began by establishing its field of action across the urban scales, from the territory, to the city, to the neighbourhood and even to the architectural building. In a bottom-up action, the local level takes the locomotive and moves across spatialities, giving each stop its own set of actions and impacts. Identifying the 'relevant' perimeter of the urban project is no easy task, given the scale of the operations involved. Nevertheless, mastering and managing the "target area" makes the action concrete and the strategy relevant. It calls for interdisciplinarity and transdisciplinarity, and this is precisely the content of the third course.

## COURSE N°III: URBAN DESIGN /// APPROACHES/PROCESSES AND DISCIPLINES



### INTRODUCTION

In order to keep up with the times, **every city** proposes an urban project. Whether it's to rehabilitate its historic centre, transform the land rents left by the relocation or closure of businesses, design a cultural or tourist complex, change its old transport network and replace it with modern infrastructure, convert to the tertiary sector, renew its housing stock... the urban project strategy is the approach to follow. There are certain constants that can be identified and certain criteria that must be met, whatever the diversity of the projects, where multiple disciplines come into play. What is it compatible with?

### III. WHAT IS THE BIRTH OF A NEW DISCIPLINE AND WHAT CAN IT BE ASSOCIATED WITH?

From the outset, several authors have associated the urban project with "urbandesign" (**INGALLINA**, 2001, **MERLIN AND CHOAY**, 1996), while others have emphasised the difference between these two approaches (**GIRALDEAU**, 1990). Some theorists locate it at "an intermediary level between planning and architecture" (**DEVILLIERS**, 1990) or between the plan and the traditional project (**BUSQUETS**, 1990).

In relation to these scales of the city, which also refer to specific disciplines, there is disagreement about the definition of the urban project and the attempt to associate this concept with an existing discipline. Others speak of "project urbanism", to describe the urban planning which took place in the city of Barcelona (**BOHIGAS**, 1985), or of a "project urban planning system" when it is "the particular projects which make up the overall plan" (**MERLIN AND CHOAY**, 1996, p. 647). Project-based urbanism refers back to the practices of the early 20th century and to the 'City Beautiful' movement (**MARSAN**, 1994, p.324), to the 'Modern Movement' (**PINSON**, 1997, p. 93; **DEVILLIERS**, 1994), and it is to distance itself from this that the term urban project has been used since 1970. This hesitation to bring the urban project closer to existing disciplines and to mark its specificity, shows that this concept is in the process of being defined (**COURCIER**, 2002).

Rather than confining the urban project to one or other of the disciplines or scales of intervention, the interest of this concept lies, on the contrary, in its capacity to link and articulate. In the words of **DEVILLIERS** (1994, p.12), "the urban project is not a profession, but a skill; it is not a procedure, but an approach accompanied by a procedure". It's about putting together a multidisciplinary team and adapting an open attitude. It's about working with what already exists". So there has to be consistency between the different levels and disciplines.

### III.1. 1. THE URBAN PROJECT AND THE DIMENSIONS HIGHLIGHTED.

Perennial constants structure the foundations on which the urban project is designed. It is confronted with a city that already exists, consolidated to use the expression of **MASBOUNGI** (2002), with an identity established over the ages, a well-established economic force even if it is inoperative, and a permanent way of life. Managing to move between these permanent features and the proposed recompositions requires the urban project to be flexible and adaptable, while at the same time working in synergy with all these features. The urban project therefore relies on its three dimensions to fulfil its mission as a planning approach that replaces traditional planning or "comprehensive planning".

### III.1.1.1. POLITICAL OR INSTITUTIONAL DIMENSION

The tools of traditional planning proved obsolete in the face of the new approach. Economic growth, driven by supply and demand, became an issue of competition between countries, cities, regions and territories. Urban planning was put to work to attract investors who would create jobs and activities and drive expansion, in order to make the competitive space as attractive as that of other competitors. It was therefore necessary to institutionalise these strategies by introducing consultation and negotiation procedures into the legal framework of the town planning instruments of the time. It was necessary to emphasise local ascendancy and coherence in a permanent "back and forth" movement. The urban project is therefore based on a bottom-up logic that enables the project to be understood in terms of its local and global problems. It is the idea that a particular project should be an opportunity to question needs at different scales, by proposing legal institutions capable of protecting and anchoring the idea in its spatial dimension. It is this local, global coherence, delimited in time and space, that characterises the political and institutional dimension of the urban project.

### III.1.2.2. ECONOMIC AND FINANCIAL DIMENSION

As a result of competition between cities, the region has become a product to be sold and marketed. Its value is calculated according to its potential and the opportunities it offers the market. Actions are then taken to attract investors, with the aim of setting up urban development projects for which sources of funding are a prerequisite for feasibility and implementation. The actions undertaken are linked together in a mechanism that guarantees the "project product". Partnerships are formed, percentages are clarified (investment at 30, 40, 50%...), strong lobbies can take the lead in execution according to their interests and redirect the project towards their own actions. This is what Philippe **GENESTIER** (1993) denounces in issue 288 of "L'Architecture d'Aujourd'hui", when he writes that *"the urban project is an operational process whose object is the city, which responds to a market logic". This state of affairs is a drift away from consultation, where influential groups impose their ideas like entrepreneurs, and the loser is the ordinary citizen. The urban project will then come under the ideology of liberalised sub-pragmatics.*

### III.1.3.3. URBAN PLANNING AND ARCHITECTURE

**BUSQUETS** (1990), stresses that the great significance of the urban project comes from its relational possibilities (articulation of spaces and buildings, confrontation of symbols and functions which lead to the deepening of its content and its justification). The town planning project has often been legitimised by external data and applied without taking into account the specific features of the place, whereas the urban project must emerge from the local dynamic (**REMY**, 1998). The traditional urban planning project is the result of an expert approach based on a logic that is intended to be scientific. It is based on general objectives defined by planners. Only when the figures in the programme, such as heights, densities, surface areas, etc., overshadow the project, the purpose of the site, its specificity and its symbolism, does it allow itself to be locked into a legal and regulatory approach, but demonstrates adaptable flexibility. It is a process of insertion and integration that offers openness to any obstacle. It also ensures :

- a land base of varying size ;
- long-term design and development ;
- the ability to control and influence the neighbourhood ;
- attention to the functioning of the new or remodelled neighbourhood ;
- a collection of economic, legal, institutional and conceptual proposals to make this project possible and feasible;
- a proposal for collective images of the city and the remodelled district.

### III.2. 2. OPENNESS AND FLEXIBILITY OF THE PLANNING PROCESS

The urban project is designed to facilitate the formation of partnerships around concrete objectives. It is not static and set in stone; it fits into the changing environment by finding its place in the urban dynamic. Its rationality is not given in advance; it is created during the planning process. It is an attitude that allows openness to opportunities and certainties. It is a stance that seeks to involve a wide range of players in the design and implementation of the project. The urban project is part of an open and flexible planning process. It enables stakeholders to learn within the planning process and the project to be taken on board by the population (**COURCIER**, 2002). It is concerned with the gradual construction of the meaning of the project. Planning must allow for fluctuations, creativity and vagueness... but on the other hand, if this management is not to turn into complete empiricism, antithetical to urban planning, the 'essential' project must be even more than in traditional planning, formalised, explicit, sustainable, totally accepted and appropriated by all investors. Strategies must be more procedural, focusing on ways of problematising and schemes of action, rather than on particular choices (**ASCHER**, 1995). The urban project proposes a conception of the relationship between knowledge and action in which there is no longer a relationship of subordination between one and the other, as in the past, but rather a continual to-and-fro between one and the other.

In the comprehensive planning model, the project is something that is well defined, or it will be carried out as planned, or with derogations. The possible gap between project and implementation is not taken into account. What is displayed is a linear vision of the design, implementation and monitoring of a project, as proposed in the work by **WRENN** (1983) on the development of waterfronts FIGURE 2 .

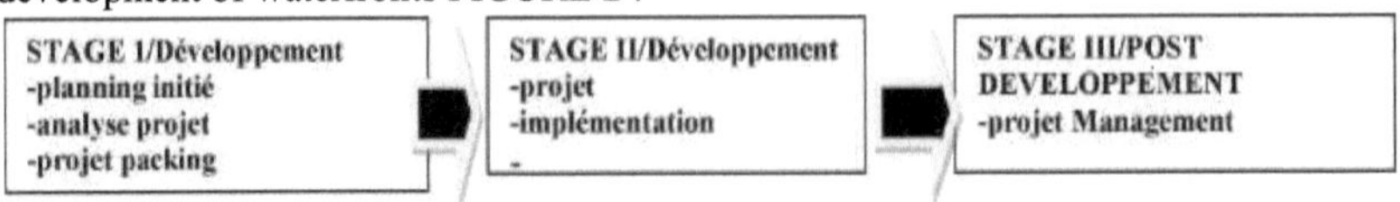


**FIGURE 2: Development process (WRENN, 1983, p.74, in Courcier, 2002, P.19)**

The urban project will evolve as it is designed and implemented. The decision-making process is seen as a pluralist process of convergence towards a compromise, as described by **MONNIER** (1992) in relation to the re-evaluation of public authority action. The programme does not progress linearly from the initial declaration of intent to the results, but is transformed under the effect of successive conflicts, and a decision emerges from the exchange of points of view and arguments in various ways. The decision-making process can be presented very schematically in the form of a whirlwind flow comprising three main components: formulation of the action, implementation of the action and evaluation of the effects. To summarise the differences between the two types, the table at summarises the dimensions put forward by traditional planning and the urban project.

**III.3. water 1. Comparison of the dimensions highlighted in traditional planning and urban projects**

| TRADITIONAL PLANNING | URBAN PROJECT |
|---|---|
| **Global coherence and top-down logic** | **Local coherence and bottom-up logic** |
| • Exhaustiveness at city level, integrated overview of the different dimensions of the area<br>• From master plan to local project | • In-depth study of a specific project limited in time and space<br>• From specific project to global plan |
| * **Instrumental and bureaucratic logic**<br>* decoupage, functionalism<br>* legitimising the project with data exterior.<br>* general public interest objectives defined by planners. The State is the decision-maker | **-Integration and democratic openness**<br>* integrating the different dimensions of the project<br>* comparing the internal components of the project<br>* specific objectives, general interest to be built up gradually with the stakeholders. Creation of partnerships |
| **-Linear planning process**<br>* clear distinction between design and production<br>* Precise, prescriptive and fixed scheme | **-open and flexible planning process**<br>* the stages of the project are consolidated<br>* open to opportunities and uncertainties |

Source: (in Courcier, 2002)

## CONCLUSION

The urban project affects multiple scales, calls on a variety of disciplines, is subdivided into dimensions according to specialities (institutional, economic or urban and spatial policy), is flexible and adaptable in terms of the actions to be taken and put into practice, adapts to the consequences on the ground, bends to the demands of reality, is integrated into processes that have already begun, takes account of general interests while avoiding friction between stakeholders, is surrounded by a democratic openness, and creates urban marketing around the project.In short, it is a bottom-up approach to traditional planning.

The urban project is ultimately a flexible planning strategy. It crosses all the scales of an area and is defined according to the relevant perimeter, involving consultation and participation by all, because it concerns everyone. Its definitions are becoming clearer, and this is the subject of the fourth lesson.

## COURSE N°IV/// URBAN PROJECT//ADDITIONAL DEFINITIONS



### INTRODUCTION

Before putting forward any definition of the urban project, we have first decomposed the content, foundations, procedures and processes that characterise this concept.

As it is a polysemic and ambiguous concept, we have had to work around it, identify it, and gradually get to grips with its various considerations, and really understand the ins and outs of this process.

We will look at some of the definitions put forward by theorists of urban planning and the basis on which they put forward this or that notion. We end the course with a definition of our own.

### IV. DEFINITIONS OF THE CONCEPT

Firstly, there was traditional top-down planning, i.e. the orders came from the central authorities and then spread to the regional and local levels. The base, which is the town, is subject to the project and called upon to carry it out, even if the project does not fit in with its development strategies. It even happens that an imposition from above forces the local authorities to accept and implement such and such a project, even if it is against the interests of the municipality. So to deal with these "constraints", which sometimes destroy the past and erase identity, a new concept emerges from the lot and imposes itself through its flexibility and its reconciliation with all the parameters.

### V. .1 SOME THEORISTS AND THEIR APPROACHES

We'll look at the opinions of the various parties and the tendencies of each. The synthesis will be made by bringing together the various constituents.

**IV.1.1 According to HAYOT (1997, p.22),** "*the urban project stems from a critique of regulatory and technical urban planning and provides an interesting alternative.* This vision is in fact global. It calls into question traditional planning, based on very strict regulations dictated by the central authorities and imposed on local populations. **HAYOT'**s alternative sees itself as an escape route from these technical obligations.

**IV.1.2 According to LAPERRIERE (1998, p.116),** he uses **STEINBACH**'s (1993) definition of urban form and adds that it is an approach which gives a central role to the players. He emphasises the process that is implemented from the island to the agglomeration.

*"The urban project is the product of a meeting of social players to define the issues, choose strategies and find ways of shaping an urban space. This meeting, which is both conflictual and cooperative, goes beyond the simple collage of objectives and visions".*

It therefore assumes the existence of :

- of a **"desire for a project"** based on the conviction that the shaping of a space can bring about changes for urban society as a whole;
- of a **"project situation"** that challenges social players by encouraging them to think, to imagine and to want change;

- a **"common referent"** for the various players, enabling them not so much to develop a consensus as to share perspectives on the issues facing urban society;
- a **"context of action and fieldwork"** which forces the players to work in real time in the places and frameworks they have built up.

It is the articulation of these 5 components - volonteer-situation-referent-permanence-**context...** that defines the urban project.

The latter bears witness to the transition from the city-as-object to the city-as-project, which is not necessarily complete.

This definition, which focuses on the procedural dimension, calls for the implementation of a project approach that is consistent with the desire for openness and integration put forward to characterise the urban project. **LAPERRIERE** speaks of a project that is the product of interaction between players, and situates it both in relation to general perspectives for urban society and a context on the ground that enables it to be anchored in everyday reality.

This definition does not reflect the idea that "the desire for a project" is not something innate. The value of initiating an urban project is far from obvious to many players. A project emerges from reflection, from a situation, from a space and above all from a genuine connection between everyone.

**IV.1.3. According to P. MERLIN and F. CHOAY (1996, p.646)**

The political urban project is a project for the city. It proposes collective images of the future. It is a way of reaching decision-makers and winning the support of the community's population through the affirmation of an identity and a shared vision of the future. "*The urban project mobilises all the players around a future image".*

Our references enable us to propose a definition of the urban project, which remains an ambiguous concept and which more or less reveals its specificity. The aim is to situate the notion of the urban project in its context, to show what this concept covers, its origins and its different meanings, and to propose a broad definition, encompassing as many references as possible, proving that it is a particular planning practice that enables action to be taken on the city through a specific project.

The notion of an urban project is a framework for thinking about how to regenerate the city for the benefit of its residents.

**IV.1.4. According to P. INGALINA (2001), S.COURCIER (2002)**

The framework of thought includes in its wake an economic factor of social and urban cohesion based on the definition of **SOKOLOFF** (2000, pp.51-61) who defines it as *"that way consisting of a flagship operation carrying an image and which must induce accelerated economic, social and urban development"*. In this sense, it is based on the city of Barcelona's project, which took twenty years to complete, making the city an open-air construction site in constant evolution. The concept is based on the problems faced by any city (transport, with the omnipresence of the car, noise, air and visual pollution, urban sprawl, the deterioration of old historic centres, insecurity, fragmentation, ....): the planning methods used up until then had proved inadequate to deal with the new problems. What was needed was a response in the form of management and intervention policies aimed at revitalising the area, based on progress, sustainable development and social expectations. A specially adapted concept was endowed with all the qualities required, including flexibility, adaptability, consultation, partnership and comprehensiveness, to cope with the new demands, and it took off, **dubbed the "urban project".**

### IV.1.5. According to ARIELLA MASBOUNGI (2002, p. 23)

The process is seen as a sum of investigation, and as an alternative to traditional planning, creating a break with the sensitive and strategic approach to the city. Its framework of investigation is both the public space, which is called upon to enable the urban project to build its organisation on that of the existing network in order to better harmonise it on all levels and territorial arrangements, articulating in time and space different dimensions of both the real (public spaces, infrastructure networks, technical networks, etc.) and the virtual (interconnected networks of service companies, commissions or associations). As a result, it becomes multidimensional and a regulator of conflict. This vision leads us to Ariella Masboungi's definition, which defends the idea of the different issues that the urban project could raise.

The concept in question is defined as follows: *"a thought-out and designed strategy for the city, an architectural and urban expression that shapes the city and addresses social, economic, urban and territorial issues".*

Although this definition emphasises urban or even territorial coherence, it does not explicitly include the consultation that was at the root of the urban project. As a founding element, the interaction between the different partners is latent. The author of these words highlights the constants that all urban projects must respect, such as sustainable development, the quality of the landscape and the natural environment, economic issues, transport infrastructure and accessibility for all. It is in fact a desired outcome, desirable but uncertain. If it is to be achieved, the urban project needs players with a shared vision. It requires a strong political commitment to combat barriers of all kinds, whether physical, functional or social. It needs the support of citizens to respect the sites and their history. It is calling for funding to enable the drawings and designs of the joint project to be realised. It is asking for new rules and new tools to match its ambitions. The urban project must therefore function with the permanence on which it is based and the flexibility associated with an ever-changing society and market.

## IV.2. URBAN PROJECT MYTHS AND DERIVATIVES

The urban project is seen as bringing together the different players and taking into account the complexity of the city. It has important advantages, but it can also be an illusion or a deception, or the use of a fashionable vocabulary.

**DEVILLIERS** (1988, p. 250) cites architects who "reject the notion of the urban project and the global utopia that may underlie it, i.e. "*urban planning proposals that claim to be capable of thinking and living an entire district".*

**BOUTINET** (1993) identifies a certain number of drifts which he describes as *"pathological", in* the face of the general infatuation with *"project management". As for* **WILLIAMSON** *(1995),* he presents a critical approach to the methods currently used for urban development, but does not recommend an alternative to these methods.

## IV.3. THE GENERAL SOLUTION

Just because a strategy used in one country with a specific context has been successful does not necessarily mean that it can be transposed.

The choice of a generic solution that does not respect the specific characteristics of the site should be avoided. The miracle project may be seductive, but the results are not very convincing. For example, the GPV (Grands Projets de Ville) tramway works and the effect of exclusive right-of-way public transport (TCSP) are taken as examples. These are major projects that bring cities together, with inexhaustible arguments and justifications, yet there are harmful and unforeseen effects that come to light later on, impacting on the life of the city. A decade later, we are faced with inconsistencies that were treated as intrinsic qualities at the start of the projects. Temporality helps spatial inconsistencies to emerge, because everything changes and nothing is constant.

**IV.4. THE URBAN PROJECT OR THE ILLUSION OF CHANGE**

The urban project, despite its intrinsic characteristics, can be superficial, focused on the final product, but does not sufficiently question the roots of the problem, the foundations of the project and the specific identity of the community. There are many examples of this, especially in Europe and the USA. The project is seen as a decoy that takes the place of the actor's imagination and becomes an autonomous object that finds its justification in itself. It ceases to be an intentional, transitional act and becomes a finite, closed, finished object, the reflection of a self-satisfied imagination (**BOUTINET**, 1993).

Cities that sign up great masters to prestigious projects think that they have carried out an urban project, but in reality they are seeking such publicity that there is every reason to believe that their primary aim is not quality and urban coherence, but media impact. This obscures the general absence of project-based urban planning and consultation with stakeholders and users. This is the case with the Grand Mosque in Algiers, the Hassan 2 Mosque, and other projects by famous architects.

**IV.5. GLOBAL DEFINITION/ OUR OWN REFLECTION**

According to theorists, the forms and content of urban projects vary greatly: urban projects often design the development of multifunctional neighbourhoods, combining housing, economic activities, the production of public spaces, shops and leisure facilities, etc. Some talk about urban projects for the restructuring of a large public space [HUBERT et al. Some people talk about urban projects for the restructuring of a large public space [**HUBERT** et al., 2017], or for a major commercial or leisure facility, such as EuropaCity in the Paris region. Researchers who have written on the 'urban project' often point this out as a prelude to their work: the expression is vague and polysemic [**BOURDIN**, 2001; **INGALLINA**, 2001; **ARAB**, 2004], which probably partly explains its success. The notion can refer to a political project, that of elected representatives, to transform a city: in this case, it is associated with a programme and local policy commitments. For architects, the term project refers above all to the work involved in designing, drawing, shaping and spatially arranging buildings and public spaces. For urban planners, the urban project refers to one or more urban operations, often major, which result in the physical development of a new district or the redevelopment of existing urban spaces. The term is also associated with teaching practices in urban planning institutes, which provide a deep-rooted structure to learning, even if they do not actually correspond to students' future professional practices. All over the world, cities have implemented urban projects - alone or in conjunction with the State and/or the private sector - to position themselves on the chessboard of national, continental and international competitiveness and territorial attractiveness. The use of this method of producing the city is part of the context of a changing urban, socio-cultural and economic system. The urban project is being implemented in a context of reconfiguration of local urban governance, marked by uncertainty, a multiplicity of players and the use of ad hoc structures. As a lever for local public action, the urban project shapes the city, whether in the way in which the actors construct the issues at stake in urban policies, in the nature of the objectives defined for them, or in the ways in which these policies are implemented (**PINSON,** 2000). We are thinking here of works based on the normative approach (**MASBOUNGI** and **DEVILLIERS1996**; **MANGIN and PANERAI** 1999), analytical and genealogical

(**HAUMONT1993**; **TOUSSAINT and ZIMMERMAN** 1998; **INGALLINA** 2001), actionist (**PINSON**, 2002, 2009), sociological (**MARTOUZET**, 2018) or indeed critical of the ideology of the project (**GENESTIER**, 1996, 2001).

**Table 2. Urban policy issues, objectives and methods**

| PRISM | CONTOURS | AUTHORS |
|---|---|---|
| INTERESTING THE SPATIALITY | -approach starting from 1'analysis physical of transformations and changes generated by the registered urban project | PANNERAI and MANGIN(1990), INGALINA (2002), MASBOUNGI (2016) |
| INVOLVING STAKEHOLDERS | -approach aimed at all the stakeholders responsible for implementing the URBAN PROJECT. | -ZIMMERMAN(2000,2004), COURCIER (2002), PINSON (2009), HUBERT et AL (2014), MARTOUZET (2018) |
| INTERESTING THE TEMPORALITY | -dynamic approach based on the transcription of the URBAN PROJECT narrative using the time taken. | ARAB(2004), GROSBELLET(2016), JOLIVERT (2018) |

Source: author, 2023

## CONCLUSION

The literature review guides us towards the urban project using several approaches. The spatial approach looks at the physical transformations of a site, generated by the actions that have been programmed and, of course, the changes that have taken place. Philippe **PANNERAI AND MANGINin** 1999, Patricia **INGALLINAin** 2002. The second approach focuses on the actors who have the opportunity to intervene in a project or action and who have decision-making power: **ZIMMERMAN** in 2000 and 2004, **PINSON G.** in 2005, 2006 and 2009, Sabine **COURCIER in** 2002, **HUBERT** et all in 2014 and **MARTOUZET** in 2018. They all assert that political decisions, the involvement of stakeholders, and their readiness to respond to hazards and resolve obstacles are the keystone of the urban project. Another approach incorporates temporality, transcribing the urban project in its entirety, and we base this on the work of **ARAB** in 2004, **GROSBELLET** in 2016 and **JOLIVET in** 2018.

In any case, the urban project has come to shape the city through major urban projects, calling on the players concerned in line with the urban policies and regulations in force and their practical implementation. Whether in practice or in teaching, the urban project is a powerful lever for public action, rallying citizens, local elected representatives, investors, bankers and users around its objectives. The actors involved in the action are important because of the political will to undertake work and to stick with it despite the obstacles encountered. According to **PINSON** and **VION** (2000), the urban project is seen as a tool for repositioning the city internationally, a desire to unite urban players in a context of international urban competition. Major urban planning projects, some of which will be discussed in the course to follow , show that urban policies are above all decisions taken by responsible and willing players.

## V. LESSON N°V/// URBAN PROJECT// POLITICAL PLAYERS AND MAJOR URBAN PLANNING PROJECTS



### INTRODUCTION

We all know that urban planning is first and foremost a political act, and depends largely on the will of those responsible for a given locality. It's an opportunity for local elected representatives to shine, especially in Europe. A number of operations have been planned and have changed the image of the revamped cities. This course focuses on a number of urban planning projects, with their aims and challenges, demonstrating the ambitions of the political players involved. The pioneering French and Italian examples of urban planning are highlighted through a number of flagship cities. The example of Barcelona 22@, EURALILLE and a few other projects will be dealt with by the learners in the second section (sub-section on foreign examples).

### V.THE URBAN PROJECT AND ITS PRACTICAL APPLICATION IN EUROPEAN CITIES

The rhythm of political deadlines (elections) pushes the political actors responsible for the destiny of towns or regions (mayors, prefects...) to put in place their own projects without necessarily taking the time to build them or to look at the merits of what they impact on. It's because we know that the urban project is seen as a way of revitalising the city, attracting investors and tourists, creating jobs and giving the city a new image and a higher ranking in the list of cities, that the political players are trying at all costs to put their projects on the agenda.

There is hardly a city that does not have its own project, especially if it has ambitions and opportunities:

-Port, industrial or military wasteland (urban regeneration), or in any case land;

- an event, - new centrality, - urban link between two poles, urban renewal...

There is no shortage of examples: infrastructure projects (LILLE, EURALILLE TGV station), tramways (Strasbourg, Bordeaux, Nantes), urban redevelopment (Bologna, Italy), innovative neighbourhoods (BARCELONA (22@Barcelona), etc.).

In cities around the world, a number of projects have been carried out to meet specific and complementary objectives. The ambitions are economic, spatial or symbolic. They have clear aims from the outset, which have been explained during consultations, supporting the words and wishes of the political players. In most cases, the results of these projects have been disappointing or even better than expected, leading to structuring impacts that run counter to the objectives set beforehand. Each project is a lesson in urban planning. Some of the urban planning projects are described in detail.

## V.1.PIONEERING URBAN PROJECT/STRUCTURING EFFECT

Each of the urban renewal, restructuring or modernisation projects carried out in neighbourhoods or even entire towns had a clear aim at the outset: to trigger an economic dynamic, change the image of the town, counter urban sprawl, reclaim the town centre or rebalance the conurbation, and so on. In most cases, these projects have had a greater impact than expected and, in some cases, have been disappointing, resulting in effects that run counter to the objectives set in advance. What do these effects mean?

As with the urban project, this concept is made up of the words "effect" and "structuring". Taken individually, they give, according to Le **PETIT LAROUSSE**(2019) :

-**effect**: the result of an action, what is produced by something. This brings us back to the word **impact**, which is defined as the effect produced by something. Looking for effects means looking at the impacts that the action has produced.

-**Structuring**: which allows structuring, favours it, determines it. To structure something, we need basic elements that play a determining role with all the components of a structure, a framework.

**Structuring effect**: having a long-term impact and bringing about structural change, encouraging coherent organisation, stimulating dynamism through large-scale action, disfiguring the initial framework, adding to it or subtracting from it.

The structuring effect of an urban project must be in some way an inductor, a lever arm and a catalyst capable of transforming, restructuring or even revitalising the area.

The structuring effect of a project can be spatial (in the case of large-scale projects, acting on the initial structure of the town), economic (in the case of an action that stimulates the market, creates jobs and attracts businesses and tourists), symbolic (in the case of an emblematic action, reinforcing an already established image and focusing attention on a specific point) or all three at once (in the case of a large-scale urban project whose aim is to revitalise, enhance and transform the spatial structure of a town or conurbation).

The structuring effect has been explained by transport specialists such as **PLASSARD** and **BONNAFOUS** (1974) on the study of the structuring effects of transport supply. The aim was to make transport policy a factor in development by *"using a mechanism for structuring space in which the transport offer goes beyond the role of a permissive condition and becomes a stimulant among others" (*.bid. p.14*)*.

The transport offer of a communication axis is considered to be a catalyst for regional development, an agent of impetus and a vector of propagation. Weber's theory of the establishment of "growth poles" linked to transport supply, creating external effects through the law of supply and demand, which encourages companies to seek the most optimal location to maximise their profits, has been studied by Bonafous **and Plassard**, who point out that in practice, the effects produced are not necessarily those expected, and the impact differs from one case to another. **OFFNER** (1993, 1995) questions the existence of a systematic effect of a transport offer on the dynamics of cities if it is not accompanied by a local dynamic involving the responsibility of the local authorities of an area as well as economic operators. If there is no real development strategy to take advantage of the transport offer (whether it be a tramway line or a metro), not only will it be of no help, but on the contrary, it risks becoming a handicap by amplifying the already existing situation. Claude **HARMELLE** (1982) explains that means of transport alone do not move the economy of a region in a strong way, but rather they are linked to production centres, and it is these production centres that call for means of transport. In short, for him, means of transport are induced rather than inducing effects.

### V .1.1 BILBAO AT THE HEART OF CONVERSION

The main reason why we have chosen this example is that this urban project combines **the qualities of an urban project** (complex for the Abandoibarra district and the buildings, local for Bilbao, conurbation for the region and national for the Basque country); new regulations, new tools, political will and strategic planning, citizen participation, flexibility, public/private partnership, enhancement of the city's image; recomposed identity and genius of places, and with **structuring effects**:spatial, economic, symbolic).

A look at the geography of the Spanish Basque Country (or Euskadi) reveals that it is one of the 17 regions or "Autonomous Communities" of the Spanish state. Situated in the north of Spain, on the western slopes of the Pyrenees and along the Bay of Biscay, this territory is home to 2.1 million inhabitants in three provinces: Vizcaya, Alava and Guipuzcoa. The administrative capital of Euskadi is Vitoria; the main city is Bilbao (population 355,000). The Guggenheim Museum is one of the most important elements of Bilbao's urban redevelopment plan, launched in 1991, which has earned the city its reputation. At the invitation of the Basque Government, the Solomon R. Guggenheim Foundation is considering Bilbao's application to host one of the institutions in its worldwide network.

The project, judged to be both unrealistic and unreasonable, was strongly criticised in the Basque Country. It was unrealistic because, at the time, Bilbao was considered an unattractive industrial city in full economic decline, with no tourist attractions and which tourist guides advised against visiting. The flourishing iron ore industry, extracted from the hills in the area and exported to England, was in decline as a result of the economic recession. Faced with competition from other European cities to attract tourists and job-creating investors, it cannot afford to pursue its policy. This is unreasonable, because the financial cost is considered exorbitant at a time when the Spanish Basque Country is going through an unprecedented economic crisis: in 1990, factories are closing one after the other, GDP growth is zero and the unemployment rate is hovering around 20% (**MASBOUNGI**, 2002). In this context, choosing culture as a means of conversion for Bilbao seemed a fool's errand. Bearing the scars of the economic crisis of the 80s, which devastated Spanish Basque industry, Bilbao began its renaissance at the dawn of the 90s. In 1989, the Spanish Basque Country launched a vast revitalisation plan with the aim of turning Bilbao into a city of advanced services in a modern, competitive industrial region.

#### V .I.1.1. Consensus strategy: overcoming conflict

Thanks to the political will of the local authorities, Bilbao wanted to revitalise its development, enhance its image and convert to the tertiary and quarter-century sectors in order to meet the demands of the time and bring itself up to date. A development plan was approved in 1994. Called PUGB (Plan Urbain General de Bilbao), its aim was to completely transform the city, especially its old districts near the river. It contained a series of large-scale projects which, according to the prophetic vision of the decision-makers at the time, would turn Bilbao into a Mecca for tourism and investment. Thanks to a coherent urban project, in a short space of time the city went from a situation of deindustrialisation and extreme urban desertification to that of a beacon of urban development and a world-famous capital city. By understanding the methods used in this project, we can draw lessons for all urban renewal projects. The coherent strategy on the scale of a metropolis and a region was the fruit of a **bold political will** (a consensus between the various players with a long-term vision of the project), a search for **investment with guarantees** (State aid, aid from the Basque authorities, contributions **from the local authorities, etc.) and a commitment to the project,** private contributions), **the signatures of renowned architects** (Gehry, Calatrava, Foster, Stern), **land opportunities** created by large-scale infrastructures and a highly determined **human factor** (the population). Thanks to the combination of these factors, the project has taken shape, taking the time allotted to ensure its duration and continuity, despite a number of setbacks during the execution of the works. Project management had to be carried out with an iron hand to achieve consensus between the various partners (**CHADOIN** et al. 2000).

**V .1.1.2. Project content: large-scale facilities and architectural quality**

In addition to the Guggenheim Museum, the centre of attraction, Bilbao has an airport, a metro station, a congress centre, several bridges, a river front, a tramway... For each of these facilities, the watchword was architectural quality. International luminaries such as Frank O. Gehry, Santiago Calatrava, Norman Foster and others were called in. Stern.

Because beyond Frank Gehry's highly publicised building, the urban regeneration project is a profound metamorphosis that the Basque capital has embarked upon. The adventure of the Guggenheim Museum is part of an urban, economic and social project, and the visit to the site and the debates that followed made it possible to gauge the stakes involved. First and foremost, the experience shows the remarkable turnaround that has been achieved in an area that was in crisis, but has now moved into the post-industrial era and become a centre of competitiveness linked to software and high technology.

This project also focused on the tertiary and quaternary sectors (the leisure and cultural industries) as major sources of employment. However, in order to attract these industries, it was necessary to change the physical environment, which was then in a state of degradation. This regeneration project is based on an urban development axis following the river. The redevelopment of industrial and port wastelands meant that the riverbanks could be given over to new leisure, commercial, residential and cultural activities. This area, known as Abandoibarra or the "International Quarter", is seen as a strategic area for the city and is the showcase for the new Bilbao. Based on the traditional values of European cities, the programme includes a mix of activities (offices, shopping centres, housing and hotels), institutions such as the European Health Agency, high-quality public spaces and prestigious facilities (the Guggenheim Museum and the Congress Centre). Linked to the Palais des Congres, the museum is at the crossroads of the river, the park and the city centre. Gehry's building, whose stated aim was to change the city's image, is part of a wider, complex urban project that is the mainstay of new public and private investment.

**V .1.1.3. Economic structuring effects**

Since the inauguration of the museum and the completion of the transport infrastructure that accompanies this conversion, such as the tramway and metro, the benefits have not waited in Bilbao - the city - and have spread to the Basque country. -Tourism has increased by 80% in Bilbao and 27% in the Basque Country (**ALVAS**, 1998). Since its inauguration in 1997, the Guggenheim Museum alone has welcomed 1.3 million visitors, thanks to its "ship-like" architecture by Frank O. Gehry and the quality and value of the works on display provided by the Guggenheim Foundation and the New York Museum. In five years (late 2002, early 2003), 5.5 million visitors have walked through the doors of the building, generating added value of €750 million. Revenue is estimated at €989.8 million, 3 times the public investment

of €154 million. This flood of visitors needs hotel rooms and residences to accommodate them. The museum has generated the construction of accommodation (between 1992 and 2002, the number of overnight stays in Bilbao doubled), creating more than 4,500 jobs and 6,000 hotel beds. In 2003, the museum's economic impact on the GDP of the Spanish Basque Country was estimated at €173 million.
In addition to the success of the establishment, the catalyst effect on the economy and tourism has also lived up to its objectives. In fact, over 8,500 articles, 60% of which have been published internationally, have been written extolling, explaining and analysing this phenomenon of reconversion, with the aim of making it applicable to cities like Bilbao that suffer from a lack of attractiveness.
Bilbao's Guggenheim Museum is an economic success story, and has also established itself as a major tourist attraction and cultural centre. Visitor numbers remain high: in 2004, the museum welcomed some 900,000 visitors. Bilbao saw record attendance, surpassing that of the seaside city of San Sebastian for the first time, with an increase of 20% on 2003 (**HECHAM ZEHIOUA B.,** 2010).

**V .1.1.4. Spatial structuring effects :**

The success of Bilbao's urban planning policy has been recognised and acclaimed internationally, so much so that the 'Bilbao effect' is of interest to many cities seeking a new image. The land is owned by a company called Ria2000, whose shareholders are the owners. The formula worked out is that these owners sell the land for development to the developers, and the funds raised are automatically invested in the Bilbao project, especially the basic infrastructure and buildings. The project's flagship facilities have led to unrestricted development, affecting half the city. The programming of a function requires changes which, to be effective, affect a part of the city, and this part, to meet the new requirements, has to transform itself and call for a different kind of development. The "domino" effect is the basis of the transformation. It is because these are large-scale projects that the spatial structuring effect is important, leading to a metamorphosis of the image of the city built by the transport infrastructure and the emblematic buildings of the project. The spatial structuring effects were such that the major works were accompanied by measures to improve the living environment by removing old factories and replacing them with modern economic zones meeting international safety standards.

**V .1.1.5. Symbolic structuring effects**

The articles written about the Bilbao project, the effective media coverage provided by the company responsible for the project, the Guggenheim Museum and the value of its contents, the effectiveness of the accompanying infrastructure, and the harmony in which the project's various facilities reign, have all been factors in the construction of its symbolic effects. In 2003, the Pfizer Foundation awarded Bilbao the prize for "healthiest Spanish city". In 2004, Bilbao was awarded the prize for the best urban project in the world, as well as the European Urban Planning Prize . Bilbao also presides over the International Association of Cities and Ports. Finally, the Convention Centre was awarded the prize for "Best Convention Centre in the World 2003" by the International Association of Convention Centres.
Bilbao's metamorphosis is not over. The city is continuing to evolve, focusing in particular on innovation and knowledge. A plan entitled "Strategy 2010: Bilbao as a global city" aims to develop initiatives and economic activities with high added value, while continuing to promote a harmonious living environment for businesses and residents. These two projects have therefore had a real impact and a considerable structuring effect. Having played a large

part in the success of the overall regeneration strategy, it is undeniable that their high architectural value has played a major role. Today, however, the construction of unique buildings is increasingly attracting media attention, and generating images that are in turn feeding into the image of the city that commissioned them.

**V .1.1.6. The Bilbao effect in Bilbao-city**

The Bilbao effect is much admired, and European cities are seeking to adapt it to their own contexts. Even in Algeria and Constantine, the idea of turning the Bardo into a futuristic business centre encompassing a museum and congress centre, which the wali (prefect) said on a national television programme would *"bring in world-renowned architects",* would be the Bilbao effect. Does it only have advantages? Of course not! Because reading what the critics are saying today about the Bilbao effect calls for some reflection. The pitfalls of a project may be latent, and will only become apparent once the project has been completed. It is very instructive to consider for a moment what the Bilbao effect has been in Bilbao itself. It is true that tourist numbers in the city have increased significantly and jobs have been created. However, the aim pursued by the local authorities initially went well beyond the tourist sphere. The main aim was to attract other types of investment, such as the headquarters of multinationals and advanced service companies. The construction of new office complexes (also with spectacular architecture) was planned with this in mind. However, the hoped-for knock-on effects did not materialise, with the result that the offices are now occupied by Basque public administrations. Another effect of the construction of the Guggenheim Museum has been the explosion in property values and prices in Bilbao, first in the city centre and then, by contagion, in the rest of the city. Housing in Bilbao is now more expensive than in Madrid or Barcelona. What's more, the wealth gap between the centre of Bilbao and the rest of the city has widened considerably. It has to be said that the local authorities draw much of their funding for their flagship projects, the Guggenheim and others, from taxes on property values, and so have a great interest in seeing prices soar (as do private speculators, of course).

**V.1.2. THE REIMS TGV STATION CENTRE (FRANCE)// WHEN THE PLAYERS ADOPT A PROACTIVE POLICY.**

As a result of globalisation and international competition, a city's attractiveness depends on the extent to which it is connected to international transport and communication systems, generating local synergies. Through these "diversified urban projects" such as multimodal stations, interchanges, airports, ports, congress centres, museums... towns are throwing themselves into the race through sophisticated urban marketing. REIMS, a town in the Marne in the Champagne-Ardenne region, in addition to its brand image personified by "champagne", has been given a project to refurbish its old station and project the TGV (high-speed train) concept, thanks to the determination of its mayor and the regional council. It's a multimodal hub combining rail, cars, urban and periurban transport, as well as the tramway. It's a bottom-up project, starting at the local level, working its way up to the conurbation and then to the national level, using this traffic system to project the Champagne brand onto the international stage. The planning strategy developed by the political players is highly flexible, the project is concerted and anticipates any potential impacts.

The redevelopment of the local brownfield sites has made it possible to embark on a large-scale project involving the neighbouring districts in an urban redevelopment that integrates the interventions broken down into "movement, functions-practices" themes, and the "micro, macro, global" scales are interlinked and can be understood as the means used to achieve the desired objectives.

The financial partnership was formed, simulations were presented, and before and after models were shown. The following table shows the impact of the TGV REIMS project.

**Table 3. Matrix approach to the Reims station project**

| Thematic issues | SCALE | | |
|---|---|---|---|
| | *Station* micro-space | *Station* macro-neighbourhood | Glob ale/linear - *city - agglo* |
| Movement | Internal organisation of the station and management of access and mode transfers | Parking and moving around the station area | Trace du TGV new station Champagne South bypass of the A4 |
| | <-Multimodal hub^ ^ Coulee verte-TCSP^ | | |
| issue | Passenger orientation | Neighbourhood accessibility | Nodalite from Reims |
| Functions and practices | Hook station-pole multimodal and service offering | Offices, housing, facilities and land management | Other spaces of inter- and intra-agglomeration development |
| | <-Shopping and services^ ^ Cultural and leisure activities ^ | | |
| issue | Information, care | Development of Clairmarais-Centre-ville | Competitive positioning agglo |
| Urban form | The station area and visual perceptions | Embedded in the surrounding fabric and support of public spaces | Sequence of train arrivals |
| | <-Public spaces, morpho | ogy^ <-Landscape elements^ | |
| Target issue' (intended audience) | Brand image and identification ^- travellers, tourists^ | Influence of the station and attractiveness of the district <-Actives, residents^ | 1st vision of the conurbation: city entrance <-Investors^ |
| Expected impact | Strong but occasional | Important and radiant | Relative but global |

Source: in these by **HECHAM ZEHIOUA BERNIA**, (2010)

This table summarises the changes that the station could bring about in the immediate area, in the conurbation and in the long term. On the three scales of approach to the TGV project, the impact is aimed at different audiences:

**-the first "*zone*" or impact ring, on the scale of the station, is** aimed at regular transport users and tourists. The aim is to provide a response in terms of image and information for immediate identification, orientation, services and shops to support travel and modal transfers. The station building should therefore be seen as a calling card. Linked to the identity of the town, the choice of architectural design should help to build the image reflected in the eyes of users, so that the management of "concentrations - dispersions" can be translated formally and aesthetically into a constrained space.

**-The second impact zone, at the level of the station area**, is aimed at residents and visitors. This zone is made up of the old centre of Reims (Haussmann-style ensemble and suburban fabric), multiple urban functions and different scales of traffic (transit, service, local residents). The target impacts relate to the attractiveness of the area and its surroundings, through the provision of appropriate services for station users. This is reflected on the ground by the provision of housing, tertiary and neighbourhood facilities, and commercial and cultural activities in an urban form that reflects the era. The district's influence is reflected in its accessibility, the quality of its connecting and supporting elements, the continuity of its public spaces and the hierarchical nature of its road network.

- **The third area of impact, on the scale of the conurbation**, is aimed at businesses and economic players. If the conurbation benefits from a facility of this scale, investors could be interested. The economic dynamic is a challenge that takes into account the quality of the landscape, the logic of viable and reliable networks, and the enhancement of local specificities in a competitive positioning vis-à-vis other conurbations on the basis of standardised criteria of service, facilities, taxation... especially as, a stone's throw from Reims, there are plans for another TGV station in the village of Bezannes, but which would have a different level of traffic. In fact, the two stations combine to accommodate all the services, but each has its own line.

The ambitions set out for the conurbation have to contend with a financial reality that has yet to be secured, and with the land freeze. The strategy of the political players is to lift the reservations made by the land registry and to find funding that could help them realise the project.

The development of the role and involvement of local authorities, which are being asked to contribute financially in order to benefit from the high-speed network, raises the question of the mutual obligations of those involved in regional development. The Reims conurbation was able to include the project as a priority in the Contrat de Plan. The State has agreed to support the urban redevelopment of a district to accommodate economic activities and alternative modes of transport that are incompatible with current land use.

- **.1.3. THE CITY OF LYON: DECEMBER LIGHTS/PERMANENT POLICY PLAN**

The festival is an instrument that gives meaning to an action, to an urban public project. It creates a part of the personality of a city, going beyond the framework of neighbourhoods or modest celebrations to take place on the scale of entire territories. In this sense, it also has an urban dimension. How does public action support, encourage or feed into this phenomenon?

A number of cities - Helsinki, Montreal, Turin and Lyon - organise a Festival of Light every year. Lyon's festival takes place in December every year and attracts a huge number of tourists (3 million in 2002, 3.5 million in 2007).

Illuminations are an integral part of the city of Lyon. Starting with tradition, every year on 8 December the city shines with a thousand lights. The "City of Light" urban project took root in the minds of the local authorities, who took advantage of the opportunity to enhance the city's image and anchor the name. As a result, the city's public buildings, historic monuments, bridges, public gardens and natural landscapes are equipped with lamps and garlands that light up at night. Each year, the theme of the illumination changes. With the help of technology, Lyonnais and tourists alike will be able to read scenes and moments from French and even world history on the facades of buildings. The aim of the lighting plan is to show people the city from a different angle, support the action taken by politicians in terms of urban planning, showcase the expertise of lighting designers and provide the city with a unique marketing plan. The effects are numerous:

- The massive influx of tourists, who need to be housed, fed and entertained, is a godsend for tourism and the city's economy;
- the attraction of investors in the lighting plan, which is boosting the sector by creating permanent and temporary jobs;
- The name is recorded in the annual list of cities that determines the place to visit in December to attend the Festival of Light;

-The effects of the Festival of Lights in Lyon are spatial, symbolic and economic in nature. The effects of the Festival of Lights in Lyon are spatial, symbolic and economic in nature. Lyon has made its mark with this plan, which is part of a series of urban projects with a structuring effect.

### V.2. LESSONS LEARNED FROM URBAN PROJECTS AROUND THE WORLD

These and other examples show that a proactive strategy implemented by decision-makers can have a major impact on the city.

- In the case of Bilbao, the writings admiring the project are eloquent, and most conclude that the city and its image have undergone a real metamorphosis. While the effects have been beneficial for the Basque country and Bilbao's image around the world, the euphoria of the first few years is now evaporating for the city itself. Who can boast of occupying an office in the renovated city? The low-income population is being pushed outwards by soaring land and property values. While the project was an urban project with a spatial structuring effect on the scale of the city, the social destructuring effect for the inhabitants of Bilbao was to
- The Reims urban project is seen as having a very significant spatial structuring effect. The particularity of this project is that it has benefited from a consensus of all the players involved, and the transformations achieved are positive. The Reims project has been enhanced at the level of the conurbation by the addition of a TGV station at Bezannes, a village considered to be a "*green bower*" by the villagers. The local authorities have warned the project's designers about the quality of the original landscape in this small community in the inner suburbs of Reims. "*We didn't want to confuse speed with haste",* said the village's mayor. The result was a surge in land and property values. Even so, the middle class moved out of the city centre, as Bilbao had done.
- In the case of Lyon, the lighting plan was a major breakthrough. Although it was based on religious tradition, evolution perfected the operation, which became firmly rooted in people's minds and became Lyon's trademark image. Anyone who mentions the word "light" automatically thinks of Lyon. In the end, this was the effect that the city's managers were looking for: to take advantage of a special event to set city projects in motion. As a result, programmes to expand, reinforce and renew the city were launched to cater for the crowds of tourists who flock to Lyon every 8 December for the pleasure of seeing the city.

#### - .2.1. THE ITALIAN EXAMPLE/MIXED STRATEGY

Theories of urban transformation serve as a catalyst for analysing the relationship between urban planning and projects, two terms at the source of a historical conflict that is now attenuated.

For Italian urban planners, urban planning is the only way of mastering the complexity of the changing city, and is itself a city project. The innovations of a number of urban planners in the field of planning within the existing legislative framework are highlighted, because they have given new impetus to the practice of urban restructuring by being part of a more general framework of strategic planning.

According to **CAGNARDI**, the urban project takes into account the superimposition of four layers:

- ***The city not to be built***. Interrupting the city's growth process means redefining an edge, giving it an identity, differentiating the city from the non-urban territory, revealing the differences, making them legible, while minimising physical growth. In addition to environmental issues, there are also questions of identity and the definition of the urban phenomenon.

- ***The city to safeguard***. Not erasing the evidence of the past, reviving significant buildings and environments, maintaining differences, not allowing one age of the city to erase the others, allowing modernity to rediscover significant, non-repetitive principles.
- ***The city to be improved*:** This is the most extensive part of the urban organism. The aim is to improve those areas where intervention is still possible, and to integrate autonomous developments into the shape of the city. A historical and unitary vision of the urban space is essential to the project, on the general scale of the plan as well as on more specific scales. Urban quality means reprojecting the urban space, defining the new city to be built, and shaping new spaces. Recovering land through conversion means increasing the amount of land available.
- ***Transforming the city*:** Brownfield sites, railway sites, etc. offer land that can be reclaimed on an urban scale. The planning of areas such as expansion zones is similar: the aim is to transform sites to improve the city, adopting innovative planning principles.

A superimposed reading of these four levels highlights the richness and complexity of the city's project.

The ***"piano regolatore"*** goes beyond the legal framework. It decomposes and recombines parts of the city, reinforcing others by improving the quality of housing. It recomposes the city within the city, directing redevelopment through major services and new parks, towards the creation of a new urban environment, places of convergence for the peripheries. The complexity of the responses to the shortcomings of traditional urban practice has given rise to more systematic and comprehensive approaches, with urban planning itself becoming the project of transformation. In this sense, the term urban project suggests new logics capable of directing the mutation of territorial configurations, referring to functional as well as physical and management variables, understood in a joint and integrated manner. The urban project is in itself a working method. Its purpose? To develop an instrument capable of governing the area as a whole. By assimilating the different scales and timescales of planning, the project considers the results from a qualitative and quantitative point of view. The urban project must also be seen as *"reading and establishing the critical distance that comes from the context"*.

According to **CESARE MACCHI CASSIA**, the significance of physical form within the urban project can be determined on three different levels:

- *the urban level*: form as an instrument for structuring the whole of a city;
- *the typological level*: form as a choice relating to parts of the town, capable of defining its morphology;
- *the stylistic level*: form as a choice of language for the city

A degree of rigidity in the structural form of the city can be compatible with the flexibility of planning procedures. It guarantees a diversity of typological and stylistic choices for building a visible urban space. The environmental quality of the city requires coherence between the three formal levels, without obscuring the importance of more pragmatic factors such as management, implementation and the definition of functions.

*"The slow times of the transformation of the urbs, with the rapid and discontinuous times of the civitas".*

Urban planners pay particular attention to the characteristics of the urban area and its physical form when drawing up the plan. The procedures used will give urban transformation operations their social and political meaning. Confrontation with society, thanks to the rules through which the various social and economic factors participate in drawing up and implementing the plan, is a central argument of democracy that the plan cannot ignore by

limiting itself to drawing alone. The rethinking of the idea of the plan, the increasing attention paid to the planning process and to the management of the plan, expresses the attitude of the urban planning and architectural project towards the city and its transformation. The coercive and peremptory character of urban plans may then fade, and a greater capacity to be modified emerges, showing the growing importance of the variable time in the historical process. The project oscillates between conservation and transformation, admitting numerous declensions.
Knowledge becomes *"an internal activity of the project itself, guiding its choices in relation to the contexts in which they will be carried out. It restores the density of quality and of the reciprocal relationships between different parts and different aspects, preventing the project from traumatically superimposing itself on stratifications and meanings. It must be conscientiously integrated into them, with a view to a slow, discreet and gradual change, which does not stand in the way of future interpretations of the traces sedimented over time".*
For **VITTORIO SPIGAI**, urban design, alongside the social and political objectives of planning, must give housing its spatial and architectural quality. ***The city as text***: the object of study, the city, is the largest, most general and omnipresent text that man has ever produced in his ceaseless intellectual and creative work;

- ***the project as research***: an applied research experience capable of verifying theories on a concrete case. On the other hand, through its intense relationship with reality, the project becomes an opportunity to generate new knowledge that will advance research. - ***three principles that determine urban success***: permanence, community, conformity. - the project must harmonise different urban scales ;
- there must be differentiation or continuity between old and new towns, the centre and the outskirts, built-up areas, adjoining areas and agricultural land;
- urban analysis and, consequently, the criteria for architectural and morphological control of the transformation.

Analysis of the visible characteristics of the city and its territory has gradually led to a new form of urban planning: third generation plans. In Italy, the dynamics of the urban project were strongly integrated into urban planning, particularly in certain experiments that began in the 1980s. In fact, the political players make decisions with the help of experts who work on the content of the project by discussing the future of the city. Nevertheless, it is the political will that takes precedence.

- **.2.2. - BOLOGNA PLAN, GIUSEPPE CAMPOS VENUTI**

It was in Bologna that reformist urban planning was most experimented with. This movement was born in the 1960s, within the centre-left political movement, and proposed urban planning reforms. In the Bologna plans of 1970 and 1976, this model affirmed the urban quality of residential areas in expanding sectors, reducing density and increasing services. The 1980 plan took on board the new developments (the end of expansion, the transformation process, the growing need for mobility, the importance of urban form, the ecological problem) and interpreted them using the "reformist key", without denying the positive achievements of the previous decades. A team of urban planners led by CAMPOS VENUTI was chosen to take the reins.
*The "piano regolatore*" innovation, presented in 1985, was an alternative to the classic policy of expansion. This transformation took the form of two types of intervention: intensive, heavier transformations on a restricted part of the existing city, and extensive, lighter transformations aimed at rehabilitating buildings in their various forms and improving the environment. The plan develops three major innovative orientations: a greater mix of

functions in the periphery, the proposal of a light metropolitan transport network, designed to support the process of decentralisation and transformation, and the organic introduction of the ecological element, through the creation of three parks, with their ecological and morphological functions.

In Bologna, the new focus on the physical form of the city establishes five morphological sectors "linked by a complex design that determines their correspondence in terms of urban and territorial morphology, mobility infrastructure, functional and environmental vocation, historical lines of economic development, and quantitative and qualitative redefinition of standards". The urban fabrics described in structural (functional, social, type of property) and super-structural (morphological) terms are a privileged field for this in-depth study.

However, between the drafting and approval of the urban plan in 1986, conflicts within the technical team and the political authorities led to the modification of this innovative instrument. The morphological component was eliminated from the urban plan and transformed into an instrument for later coordination (concerted urban design), concerning the formation of local development plans for interstitial transformation zones.

Despite this drawback, the *"piano regolatore"* of 1986 is nevertheless an innovative instrument that is quite exceptional in the Italian panorama, a solid "reformist plan for transformation". In this way, the complex urban programmes provided an opportunity to launch a new planning model, based on a logic that starts with the particular and extends to the general, where an overall vision, a "project" for the city, is codified. Operations with a high potential for transformation in contrast with stagnant or declining sites, out of balance with the general objectives for the territory as a whole, are banned, seeking a balance between investment in degraded sites and future innovation so that the city is capable of applying an urban strategy.

The experience of these programmes has demonstrated the possibility of a new relationship between planning, which by its very nature covers the whole territory, and the programme, which is limited in time, determined and endowed with funding which is already available. In the various proposals for national framework laws, there are two distinct levels of planning. The first, which is long term, sets out the major invariants of the territory and the second, which is coherent with the first, is made up of one or more short term components. These define the operational elements of the first level as common urban planning instruments.

### V.2.3. THE CITY OF NAPLES AND THE URBACT STRATEGY: A BLEND OF REFLECTION AND MANAGEMENT

Europe has initiated innovative urban regeneration policies for its cities, and is committed to building a new shared strategy based on a common culture and good governance, bringing together **local government players**, the local population and the various technical bodies. The major challenge is to boost economic attractiveness and the movement of capital. This policy has resulted in major recovery and development operations by applying this wide-ranging programme to degraded urban areas, ensuring feasibility and introducing this management of initiatives into sustainable development. Since 2002, URBACT has been the European territorial cooperation programme designed to promote integrated and sustainable urban development in the cities of the Member States of the European Union, Norway and Switzerland. It is a cohesion policy instrument, financed by the ERDF (European Regional Development Fund) and the Member States, for exchanges and learning between cities, with the aim of developing solutions to major urban challenges. By networking, European towns and cities strengthen their skills and capitalise on good practice, while supporting public decision-makers and those working on the ground, including local residents, to develop sustainable solutions that integrate the economic, social and environmental dimensions of urban development.

The **URBACT** strategy is part of a vast programme called the **"Community Initiative Programme". URBACT is based on the creation of a network of European cities** and, in practical terms, offers a satisfactory working framework covering several themes according to the needs of the cities that have joined the network. The European Union's budget is 50% co-financed, with a proposal for assistance from experts chosen by the cities themselves.

The study of examples and site visits were necessary to set up a business model structured by relevant parameters. Namely:

**- The physical and urban planning aspects of the projects**

**- The socio-economic and environmental aspects of regeneration**

**- Governance and implementation procedures.**

These three dimensions were the subject of the creation of a **specific grid** grouping together a variety of situations and scenarios. This grid is intended to serve as a guide for all urban regeneration projects, in line with priority objectives such as the exploitation of the various cross-sectoral analyses, intrinsic knowledge of the context and the identification of innovative practices to be used.

The work has given rise to economic appeals, pushing the concept further towards using this urban regeneration for tourism, which is characterised by constant growth and generates a turnover estimated in billions of euros or dollars. Before it was created, there were small programmes called URBAN. Some cities joined them.

URBACT is structured around four main objectives.

1-- the development of a structured exchange of good practice and experience between cities through the creation of thematic networks and working groups ;

2-- capitalising on and disseminating the lessons learned from the URBAN programmes ;

3-- improving the capacity for action of the players concerned, assisted by the residents in a participative approach, and therefore of the cities, in the various areas that characterise integrated regeneration policies.

As a business model, the URBACT programme is based on three indicators:

- competitiveness,

-economic growth

- job creation and community involvement

This strategy, defined in Lisbon and Gothenburg, is in line with the objectives of sustainable development and is based on the construction and management of urban development plans/projects involving the following innovations:

- the local action plan must be drawn up within the framework of the theme defined (in advance);
- the local support group must be involved in the network's activities, and in particular in drawing up the plans;
- greater synergy between the various vertical levels ;
- the participation of the Managing Authorities of the Structural Funds (represented by the regions, in particular for the preparation of the local action plan, the implementation of which could be financed by European funds.

This programme covers the following areas.

- ***Priority axis 1 - Cities, Engines of Growth and Jobs*** as sub-themes: promoting entrepreneurship; improving innovation and the knowledge economy; employment and human capital.

- ***Priority Axis 2 - Attractive and Cohesive Cities***, with the following sub-themes: integrated development of urban areas in difficulty or likely to become so; social inclusion; environment; governance and urban planning. In terms of its content, objectives and stakeholder participation, URBACT can be likened to a business model.

**URBACT and the city of NAPLES.**

A city in south-west Italy, 230 km from the capital Rome, a coastal city on the northern shore of the Mediterranean Sea, it has always been an important crossroads, providing a network of links between several regions and countries. It has been a political, economic and social centre, with a large population and numerous industries, earning it a place among Italy's top ten cities.

At a certain point, its development came to a halt. This was due to social problems, with the unemployment rate approaching 25% of the working population. The occult groups that ruled the town tarnished its image.

There have of course been development projects such as the creation of business districts, the UNESCO classification of the city's historic centre, infrastructure modernisation work, etc., but the absence of a clear policy based on good governance has left NAPLES a city prey to dubious transactions.

In the interests of a common culture, the European Union has chosen NAPLES as a pioneering city for the introduction of new initiatives, where an action plan or business model had to be put in place.

A number of objectives were set: firstly, to restore Naples' image to the outside world, which had been badly lost and poorly viewed, scaring off tourists and investors, and secondly, to re-establish 'normality' by giving the city council and local authorities a central role in managing urban change (**MOLLURA**, 2011).

The city of Naples, with a view to enhancing the city/port interface, has called on experts to design a busness model that is appropriate to its concerns. These university experts and practitioners travelled around European cities to strike a balance between scientific knowledge and the management of relationships between partners. This programme concerns cities that have benefited from the URBAN I programmes, cities with a population of more than 20,000, and cities in countries covered by Europe's neighbourhood policy that have ultimately had access to the information. The overall aim is to encourage comparisons between different cultures and thus strengthen the common culture. The URBACT programme has mobilised 20 thematic networks, 8 working groups, 3 study groups, and other initiatives aimed mainly at helping the new Member States ("Knowledge Transfer Kit", "Support for Cities") and testing the new URBACT II programme (two "Fast Track Network" pilot projects). Six themes were chosen because they were of major interest to the cities: economic activities (SMEs, innovation, etc.) and professional integration; integrated approach and urban regeneration; participation of residents; population of foreign origin; urban safety; young people: from exclusion to integration.

Several concepts were named.

- **PARTECIPANDO** (Citizen participation in integrated urban regeneration programmes as a key to social cohesion). An explanation of the concept can be found here * The URBACT PARTECIPANDO project is a network of 22 cities from 7 different European countries, led by the Municipality of Rome. It aims to capitalise on and disseminate experiences based on citizen participation in urban regeneration policies and practices. The Partecipando project's approach is based on local surveys and thematic workshops to produce valid tools for raising citizens' awareness of urban issues. The aim of the Partecipando network is to focus on concrete experiences implemented to improve the management of future integrated urban policies. The network aims to :

- creating broader local democratic policies ;
- to develop a strategic vision of local development geared towards environmental, social, economic and political sustainability ;
- improving social networks, solidarity and justice, and recognising the value of cultural and natural heritage.

Citizens can play a strategic role in strengthening the sustainable territorial development of their cities. In recent years, European cities (particularly those in disadvantaged areas) have experienced a crisis in capacity building and citizen participation. In this context, the decision-making process has come up against a number of problems. The methodological approach adopted by the URBACT Partecipando network to improve the quality of urban management experiences is based on the principle that citizen participation is a complex process whose achievement is closely linked to social factors (https://ec.europa.eu).

- **CULTURE** (Cultural activities and creative industries as key to urban regeneration)
- **C.H.O.R.U.S.** (Regeneration of urban centres through the enhancement of cultural heritage)
- **REGENERANDO** (Economic activity and professional integration)
- **GOOD PRACTICE AND "GOVERNANCE"** (involvement of the municipality as a strong link).

**Thematic networks and working groups are** the tools for exchanging experience and learning.

The other new feature is the European Commission's initiative to select certain *Fast Track Label* projects for direct monitoring of all network activities.

All cities in the Member States can be partners in the Programme, as can Regions, Member States, Universities and Research Centres active in the field of sustainable urban development. Other cities may participate at their own expense.

The Municipality of Naples has become a partner in the Programme through two approved projects:

- as leader of the **CTUR** thematic network (*Cruise Traffic & Urban Regeneration of City - Port Heritage as a key for the sustainableeconomic, social and urbandevelopment / Tourisme de Croisiere et Regeneration urbaine du patrimoine portuaire urbaine en tant que moteurs du développement economique et social*). The project studies the interactions between cruise tourism traffic and the regeneration of urban and port heritage as an opportunity for the sustainable economic, social and urban development of cities. In addition to Naples, the other cities taking part in this thematic network are : Varna (Bulgaria), Alicante (Spain), Rhodes (Greece), Matosinhos (Portugal), Rostock (Germany), Helsinki (Finland), Dublin (Ireland), Trieste, the Port Authorities of Naples and Matosinhos-Porto (Portugal) and the Region of Valencia (Spain, Valencian Community);
- as a partner in the **HerO** thematic network *(Heritage as Opportunity Sustainable Management-Strategies for Vital Historic Urban Landscapes*). In addition to Naples, the project partners include the lead city of Regensburg (Germany), the cities of Graz (Austria), Vilnius (Lithuania), Sighisoara (Romania), Valletta (Malta), Lublin (Poland), Poitiers

(France) and Liverpool (England). The aim of the project is to develop integrated management systems for historic urban centres and to disseminate innovative good practice with a view to striking the right balance between the conservation of historic urban heritage and sustainable development to enhance the attractiveness and vitality of the areas concerned and promote the socio-economic progress of the resident population. The project has been awarded the Fast Track Label.

We can say that URBACT II represents a great opportunity for :

- real dissemination of results at local level
- better communication between the Region, the State and the European Union concerning the activities carried out by the cities as part of the Programme in order to improve the use of the Structural Funds earmarked for shared projects during the proposal phase;

the construction, in terms of human resources, of a stable network for dialogue between the European Commission and municipalities: to provide up-to-date and direct "bottom-up" information on the state of Europe's towns and cities, and to help local administrations transform themselves in terms of governance;

- Promoting dialogue between the European Commission and local politicians;
- finalise the exchange of knowledge and good practice through the concrete implementation of local action plans/projects;

URBACT II has encouraged the introduction of good governance practices and the dissemination of a certain European culture of integrated urban development policies, shared by a growing number of institutional players, which should make projects more effective.

- From a political point of view, the City Council, adopting the organisational model of the URBAN Programme, has decided that, for URBACT projects, an assessor will be delegated, according to the specific theme of each network, for the cross-sectoral coordination of activities. This cross-sectoral coordination of political responsibility is reflected in the cross-sectoral organisation of technical and scientific action through theURBACT Inter-Directoral Unit.

-In addition, activities at international level (workshops) enabled the various representatives of the administration who took part as study project leaders to capitalise on the experiences of the entire network.

-All of this is fully in line with the Commission's intention, through URBACT, to have an impact on European administrations with the aim of improving cities' capacity for action in integrated urban regeneration policies. URBACT is thus proving to be an important route for training public administrators, helping to strengthen a European identity from below.

The URBAN and URBACT programmes have been the subject of comparisons between cities, becoming a common heritage in the language of local administrators: *integrated approach, transversality, project work, governance, partnership, real consultation with residents, bottom-up logic, anchoring in the territory, participation of residents, capitalisation and dissemination of results, ex ante, in itinere, ex post;* but also in relation to key words that bring local administrations closer to Europe, such as: *European Commission, management of European funds.*

**Table 4: URBACT OUT OF NAPLES (Italy)**

| **Partners key res** | **Key activities** | **The value proposition (added value/offer) is 50% co-financed by the** | **customer/decision-maker objectives** | **Customer segments/** |
|---|---|---|---|---|

| | | EU. | | |
|---|---|---|---|---|
| - | Regeneration | -Introductory programme | -insertion | -population |
| L'Union | urban redevelopment | Community URBACT | professional *- | young people |
| Europee | urban | I and II containing sub | integrated approach | participation |
| nne | -Attractiveness of towns | plans like URBAN | -capitalization of | citizen |
| -The | European | 1&2 | experiences | **PARTECIPAN** |
| cities | -development of | -Aspects | of the entire | **DO***: key for |
| european | historical centres | socio-economic and | european network | social cohesion |
| nes | urban development and | environmental | -positive impact on | -C.H.O.R.U.S |
| -the | good practice | regeneration | the administration | (regeneration of |
| regions | -improving | -Governance and procedures | in the | urban centres a |
| lesmunic | capacity between | of implementation/ | policies | Through the |
| ipalites. | directors | SPECIFIC GRID | upgrading | valuation |
| All | --creation of 20 networks | OR GUIDE TO | urban | cultural) |
| cities | themes | REGENERATION in | - constitution of | -GOOD |
| States | -8 working groups | Development | NETWORKS : | PRACTICES/GO |
| members | - 3 study groups | Durable. | City/Region/ | UVERNANCE |
| from | -economic activity | GOAL //competitiveness, | State/Union | (participation |
| Program | SMEs/Innovation. -the plan | economic growth | Europeenne... | concrete of |
| In addition to universities and research centres, it is also a framework for sustainable urban development. | local action group* - local support group - synergy between the different vertical levels - involvement of the Structural Funds Managing Authorities. Axis 1 - ***Cities, Engines of Growth and*** Jobs ***Axis 2 - Attractive and Cohesive Cities*** The European Commission's initiative and selection of certain *Fast Track Label* projects, | employment . -Transcription of an urban regeneration strategy -impact on the populations of european cities in neighbouring countries with similar policies | training for managers - common culture - similar vocabulary and concepts | municipalities and regions) - ECHELLES ASCENDANTES ET DESCENDANT ES /penser global/agir local |

Source: author, 2023

## V.3.URBAN PROJECT, TERRITORIAL SOLIDARITY AND THE CITY'S DESIRED IMAGE

The urban project, whether for a district, a conurbation or an urban region, is always a collective construction that depends on the system of players that produced it and the values they share. If the image is not to be "blurred" on the outside, it must express the cohesion of the area and "sublimate" the conflicts that may arise during projection, consultation or production. Mobilising all the players in an area around a shared development project is the essential prerequisite for building its image. Involving all the key players in the area, it will embody the solidarity that unites, around the local authorities, the area's residents, businesses and creators (negotiation, consultation and local democracy). It gains a great deal by integrating the requirements of "sustainable development" at a very early stage: a sustainable city evokes a modernity where economic development is not achieved at the expense of the living environment or social balance, and where local governance enables the construction of shared projects that respect these imperatives. The political ownership of the project varies according to the nature and size of the project: neighbourhood projects, city projects, conurbation projects, pavs projects.... Each project-building process has its own rules, procedures and ways of doing things. All of them are essential if we are to change the image of an area in any significant way. The methods used to produce and validate the project will require "project governance" involving residents, associations, economic players and public authorities in forums for dialogue and mobilisation, where the necessary consensus can be reached.

The urban project is a response to the limitations of traditional planning in a context that has changed both in terms of era and tools. Today's planning context is characterised by urban regeneration and operations to revitalise the city. The concept of the urban project reflects an approach rooted in the reality of an area and its stakeholders, and seeks to integrate and articulate different dimensions, disciplines and scales of intervention. It is part of an open, flexible planning process that can adapt to a changing environment. Indeed, parts of the city that have not been thought through can easily be transformed because of the flexibility of the process. The process is characterised by flexibility and adaptability, with a multiplicity of players involved in the planning process. It appears to be both a critique and a substitute, a method and a series of approaches, touching on the transversal and the vertical. In short, as **MASBOUNGI** (2002, p.31) points out, *"it is a methodological challenge to invent the methods and tools for moving from intention to realisation".* One of the major ambitions of the urban project is to make the city more dynamic, to play a structuring role. The cases of Bilbao, Reims and Lyon were chosen with this in mind. These projects could also lead to negative impacts that were hidden and which emerge over the long term. It is the players and their project strategy that are responsible for the effects. The specificity of our course focuses on the dual challenge of the urban project and the structuring effect, the effect of making the most of a city's opportunities, which is situated in urban marketing and which endeavours to diagnose the area, to identify its strengths and weaknesses and to dissect the threats and environmental opportunities that are offered or stand in its way. Strategic orientations should follow, leading to action programmes in the form of marketing plans. It makes residents aware of the opportunities offered by their area and encourages them to get involved in order to bring about endogenous development. It can also develop a specific brand image in relation to its environment in order to strengthen its competitive positioning and encourage external development (influx of tourists and foreign investors). It raises awareness of the factors underpinning the region's image, customer expectations and the criteria they use to choose destinations. This concept shows the importance of examining the content in greater depth from a methodological perspective. These concepts of urban project and structuring effects are compatible with the notion of brand image, since they bear witness to the consultation and opinions of local residents, which can encourage the emergence of a particular dynamic. For a project to have a structuring effect on its area, it must not only fit into a certain number of correspondences between the context and the strategies of the players, but it must also be well targeted, and emanate from a diagnosis of this area.

## CONCLUSION

In traditional towns and cities, the small size of the area and the relative simplicity of the decision-making process mean that the image of the area is perceived in the same way. This is no longer necessarily the case in today's enlarged, fragmented and compartmentalised urban spaces. Methods of representation and decision-making processes must adapt to this new situation. Producing a project on the scale of the urban region will therefore require an 'alliance' between the conurbation and the surrounding periurban municipalities. Setting up a development council that brings together all the driving forces in an area is very important to ensure the success of a project. These "governance" issues are not incidental, as they create the conditions for collective ownership of the project and the coherent projection of its image to the outside world. At other levels, cooperation is essential. Cities are never alone in the world; they are all involved in dynamics of dependence and complementarity with other cities. City-to-city cooperation and the development of their respective assets, within the framework of "city networks", are often the key to mutual attractiveness. Alliances and networking are essential. Nor can cooperation strategies between regions ignore the North-South dimension of their relational system. The circle of relationships is linked to history, to its population, and therefore to its very image. Appropriate political support will be needed to ensure that the proposed forms of cooperation are successfully implemented and taken on board by their residents. This is how the urban project could succeed. A tertiary economy, focused on services, information, the quarter, software, etc., has developed with the rise of culture, leisure, urban tourism, etc. These socio-economic transformations of the productive system have an impact on the city by modifying space and use, scales and territories, pushing the introduction into a new post industrial urban age and a new urban revolution. This is the content of course VI.

**V I. COURSE VI/// URBAN PROJECT// CONCURRENT OR TRANSACTIONAL TOWN PLANNING**

**INTRODUCTION**

**V I. TOWARDS TRANSACTIONAL URBAN PLANNING**

**V I.1. REMINDER OF PAST PRACTICES**

**V I.2. THE EMERGENCE AND GROWTH OF PUBLIC DEBATE**

**V I.3. TRANSACTIONAL PLANNING//GOALS AND NAMES**

**VI.3.1. DESIGNATIONS ACCOMPANYING THESE ACTIONS**

**VI.3.2. THE MIDDLE WAY OR THE THIRD WAY CONCLUSION**

**INTRODUCTION**

Several town planners have asked themselves the following question: where is town planning heading? How is its practice evolving? What impact is post-industrial society having on the city? On its form? On its mode of production? What new solutions have been devised? What innovative urban planning discourse has been produced? What is the value of the notion of the urban project? Hasn't the urban project become like salt in the kitchen, being used in all sorts of ways?

With the socio-economic and technological changes of recent years, the acceleration of mobility, the explosion of NICTs, the accentuation of inequalities, the process of metropolisation linked to globalisation....the city of the industrial era has been overtaken not only in its form but also in its functioning, and the tools of planning have become obsolete and definitively outdated.

The urban economy is no longer simply physical and material capital, heritage is no longer limited to historic buildings (monuments, old fabrics...), human heritage (education, health, culture) scientific (knowledge, know-how, university structure, research centre) environmental (climate, nature, landscape), all the infrastructures, facilities and specific services accumulated by a city throughout history, constituting its productive potential. The post-industrial economic system then becomes the container in which the metropolitan area is embedded in a controllable osmosis. New concepts have appeared under the names of "urban renewal", "sustainable city" and "GPV" in a city policy framework where urban planning discourse supports the trend. The concepts of technotopia, technolatry (salvation through technology), hypermodernity (urban space controlled by the dominant technician), post-modernism, ecologism (a return to naturalism questioning the inhuman expansion of environmentally destructive metropolises that threaten the ecosystem). In the face of all this, the urban project appears to be the solution, where the players are committed to it through the irruption of public debate and consultation.

- **I. TOWARDS TRANSACTIONAL URBAN PLANNING**
- **I.1. REMINDER OF PAST PRACTICES**

Since 1982, with the advent of decentralisation and Europe, urban planning has ceased to be a monopoly of the State. Competence is being sought at local level, resulting in a multiplicity of players (elected representatives, developers, project managers, financial experts, residents, etc.) and creating new professions (urban professionals) with complex interaction between partners.

The management model has been superseded by a new form of interactive organisation based on 4 main functions:

- **strategic definition**: bringing together the urban policy followed, the precise and strategic orientations, the city project as such. involving elected representatives, experts, urban planners and citizens.
- **Political support**: financial responsibility, long-term responsibility on the part of the institutions, continuity over time, periodic evaluation (elected representatives, institutions, project owner);
- **Technical management**: management and coordination of the project by a steering committee acting as an orchestra conductor, managing the translation of the design onto the ground, checking overall coherence and guaranteeing economic feasibility by bringing together developers, town planners and private and public partners,
- **- The implementation function**: the final operation to make real objectives a reality on the ground, involving operators, developers, promoters and architects.

These four functions, with their specific players, become the backbone of the URBAN PROJECT, as before it was the linear function: upstream and downstream, and since then it is ultimately the interaction (feedback) of these elements that determines the quality of the project.

**VI.2. THE EMERGENCE AND GROWTH OF PUBLIC DEBATE**

Elsewhere, at both national and local level, development projects are becoming a social issue. Public debate is introduced (local democracy, governance, representation, citizen participation in the production of space), and becomes part of the jargon of the urban project. The issue is a societal one, and it's the user who has to bear the brunt of the impact, so he becomes a player in the decision-making process. The concepts of hybrid forum, citizens' conference and consultation are becoming the leitmotiv of urban planning initiatives.

Researchers specialising in this field refer to the "NIMBY syndrome" (not in my back yard), because public representation is monopolised by notables defending their own interests, and it is no longer the average citizen who speaks.) Despite this lobbying, decision and codecision take precedence.

Nonetheless, these changes in the decision-making mechanisms are transforming old practices and, on the whole, they are having a positive effect, with the opening up of projects that were previously opaque, but with public debate, citizens are being informed of the transformations that are to come to their area. Thanks to this openness, the future of urban planning and its practice has been impacted.

**VI.3. TRANSACTIONAL TOWN PLANNING// AIMS AND NAMES**

The changes that have taken place are based on a new project culture (a project broken down into small, local, interconnected projects that need to be coordinated, etc.), a multitude of players (confrontation of opposing ideas, distinct objectives, a form that needs to be arbitrated) and a public debate on the proposed developments, involving citizens (residents and users) in seeing and being concerned about the future of the area.

These new production conditions are giving rise to a new practice known as **transactional urban planning**. Its aim is to: -organise exchanges between players ;

-coordinate actions at different levels and between different players ;

-seeking practical compromises between opposing positions and visions through negotiation;

-arbitrate in conflict situations ;

-debate with citizen users.

These main or transactional tasks are carried out in two ways:

-polemics: power struggles between players, legal action, demonstrations, occupation of premises.

- Contractual: giving priority to pre-established negotiations, committees of wise men, councils, commissions, dialogue through debate...

**VI .3.1. THE NAMES THAT ACCOMPANY THESE ACTIONS**

a) This type of urban planning is also called "**concurrent urban planning**", advocated by a number of theorists such as **F. ASHER** in response to the questions raised by the collective production of space or co-production of projects. It is, in fact, "a form of urban planning based on negotiation as part of the problem of governance". Its distinctive feature is that it acts both horizontally and vertically.

Horizontally: co-design, co-decision between private and public players and partners;

-Vertically: intervention by players in successive layers from global to local "think global, act local".

b)*This urban planning is called **procedural** because it uses rules of law. In each country, we need to look at the strategies that have been adopted and debated in parliamentary institutions. In France, for example, there are the DEFFERE laws on decentralisation, the CHEVENEMENT law on inter-community cooperation, the VOYNET law on countries, and the SRU law on diversity and participation (Solidarite et Renouvellement Urbain).

c)*This urban planning **is managerial** because the city is considered to be a business, with the main person in charge becoming the manager with, of course, the dominant criteria being fiscal profit and property gains.

d)*This urban planning is **transactional** and presupposes, from the outset, a pluralism of urban representations linked to the different players and their conceptions of the "good city", on the basis of which compromises and syntheses are constructed through negotiation.

**VII 3.2. THE MIDDLE WAY OR THE THIRD WAY**

Today's theorists are looking for a third way between the functionalist urbanism of the previous period, the modern movement with its formal models (CIAM, garden cities, etc.) and the managerial urbanism of the current period, which aims to be democratic but retains a hierarchy in the decision-making system.

There is a middle way among the dreams of urban planners, a way that lies within the history of urban planning, and which questions its political aims, such as space and its meanings, democratic society and its contributions to the discipline, and the essential link between the social project and the spatial project. This middle way is a scientific challenge to be met by combining all the ideas as 'theoretical supports' and the realisations already put into practice on site as 'executed practices'.

Local governance is essential, and delegating decisions to local authorities can, with the help of everyone (citizens, communities, associations, etc.), lead to a balanced and feasible management plan.

The diagram (FIGURE 3.) summarises the actions of each player and the interaction between them.

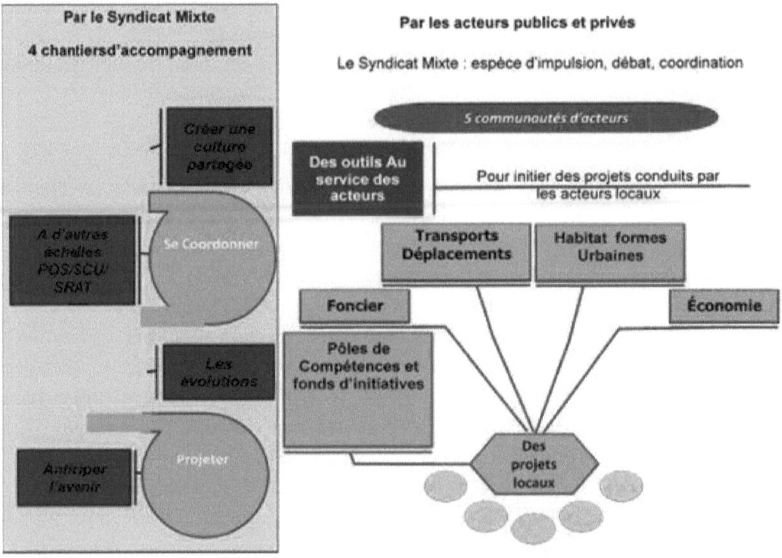


**FIGURE 3.** Actors and actions for successful local projects (author, 2023)

## CONCLUSION

Transactional urban planning is a way of tackling the problems of today's post-industrial city. It brings together all the players, with their prerogatives, duties and rights. It has to cope with the changes in cities, which are moving abruptly from the tertiary sector to the quarter-century, with softwares and NICTs. Each of the partners is bound by moral contracts that require them to put in place the tools needed to make any project a success. Since 1995, over a hundred countries from all over the world have come together **every year at the COP** to discuss the climate and the **fight against global warming**. Because of the complexity of the subject and the stakes involved, it's a long process in which environmental, economic, social and diplomatic issues come together.

# COURSE VII. SUSTAINABLE DEVELOPMENT - A STRATEGY FOR A SUSTAINABLE CITY?

## INTRODUCTION

## VIII THE ORIGINS OF SUSTAINABLE DEVELOPMENT

## INTRODUCTION

We have seen from the other courses that the urban project requires sustainable, effective and efficient planning strategies, but can encounter obstacles of various kinds. The city as a sustainable entity is concerned with the preservation of environmental, social, economic and landscape dimensions, and tends towards urban planning that reconciles all of these. After several congresses and meetings between scientists and political decision-makers, the outcome of these meetings has taken the form of charters setting out the fields of intervention and the clauses inherent in them. The change in the world's systems of government is proving to be essential if we are to make the transition to sustainable development and contribute to making cities more receptive to the objectives of sustainable development set out in the Paris Agreement and the new UN Sustainable Development Goals (SDGs).Through the content of this course, we will look at the birth of sustainable development, the objectives set, the decisions of the Conferences of the Parties (COPs) and the strategies for managing the sustainable city.

## VII. THE ORIGINS OF SUSTAINABLE DEVELOPMENT.

There are dates that mark important births in the life of a living organism, and the city is one of them. So we'll be looking at the BRUNDLAND report, the issues and objectives of the various climate conferences managing the new developments that have consequences for cities, the broken promises and the charters for action to be taken.

### VII.1 The BRUNDLAND REPORT AND THE DATES OF SUSTAINABLE DEVELOPMENT.

**The concept of** sustainable development was born in 1987 when the Brundtland Report was published. **Sustainable development" (WCED**, 1987, p. 8) aims to reconcile economic development with environmental protection in an effective strategy, in order to pass on to future generations a world that is well equipped to support and benefit from this development. Respect for the environment, social equity and economic profitability are inseparable and serve as benchmarks for all action. The "*new era of economic growth*" has given rise to a variety of interpretations, including the need to make long-term adjustments to reconcile the different dimensions - environmental, social, economic, landscape and aesthetic - with the city as the receptacle for this sustainable development.

In 1994, in the Aalborg Charter, where more than two hundred European cities signed agreements committing themselves to promoting sustainable development of their territory, we talk about the sustainable city and its specificities... The phases in the implementation of sustainable development are set out in Figure 4, showing the birth and evolution of the concept.

Key dates in sustainable development

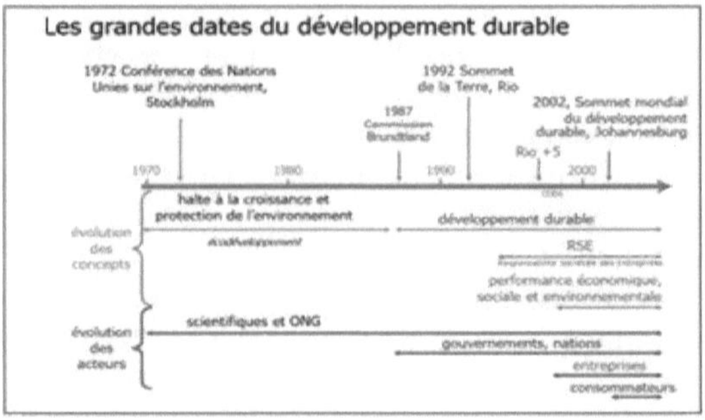


**Figure 4:** Schema of the key dates in sustainable development. Source :Christian BRODHAG, Le developpement durable du concept a l'action : le SD 21000 Ecole Nationale Superieure des Mines de Saint-Etienne, 2004

In 1992, over 178 countries met in Rio de Janeiro for the **UN'**s ten-yearly conference on the environment and development. Significant advances were made, such as the signing of the **Rio de Janeiro Declaration** on Environment and Development, which gave an "official" definition of sustainable development.

This declaration sets out a **climate convention** that supports the need to reduce greenhouse gas emissions in order to minimise the human impact of climate change (**Table 5**. COPs, year, cities, countries, objectives and actions IN APPENDICES),

The wheels are now in motion: every year, the countries that signed up to the climate convention at the **Rio summit** will come together to find concrete solutions to combat climate change: this is the famous **"Conference of the Parties"**, which will take place in a different city every year. Non-governmental players (NGOs, businesses, cities, citizens, etc.) will also be taking part, to ensure the best possible representation of society in the fight against climate change.

In 1992, the first "Earth Summit" was held in Rio de Janeiro. The heads of state present adopted "Agenda 21", comprising 27 essential principles (later 30) to make sustainable development a reality. The Kyoto Conference in 1997 saw the adoption of a treaty to reduce greenhouse gas emissions, with binding targets, which definitively established global warming as a major international concern (**BRUNEL, S.**, 2004, p. 45).

**VII.2. SUSTAINABLE DEVELOPMENT OBJECTIVES: SDO**

The objectives of sustainable development are part of a strategy to strike a balance between society, the economy, the environment and governance, "*objectives of social equity, economic efficiency and environmental improvement, which also aim to simplify organisation and democratise decision-making, management and control (governance)*" (**VALDIEU**, 2011). They can be summarised as follows:

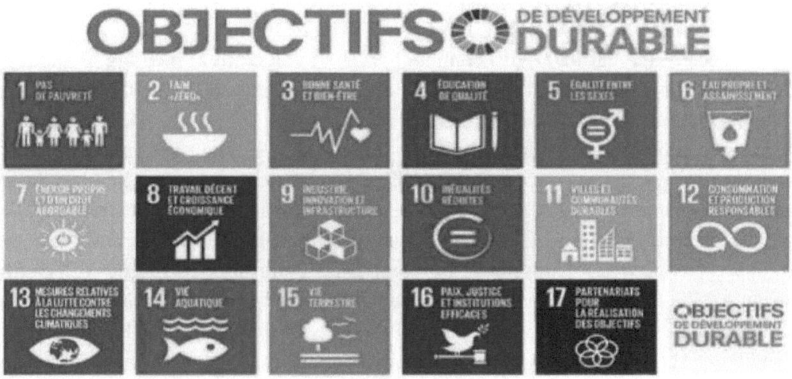


**Figure 5:** Sustainable development goals set by the UN (VALDIEU, 2011).

- Ensuring diversity of land use and facilitating the integration of populations;
- Enhancing our heritage and saving resources;
- Ensuring public health by organising the management of territories in an efficient local democracy and promoting economic growth that takes account of the environment and social equity.
- Respecting the principles of efficiency in all sectors, assessing the impact of choices with reversibility and innovating management practices;

-Think global, act local, because SD is everybody's business;

-involve the population, the users who are the primary users of towns and cities, in any concerted action (Concerted Urban Project)

*"The city can be a relevant scale for applying the principles of sustainable development, not least because it concentrates a large number of activities and people in a small area"* (**POUYANNE**, 2004). Figure 6 shows the pillars of the Sustainable Development strategy

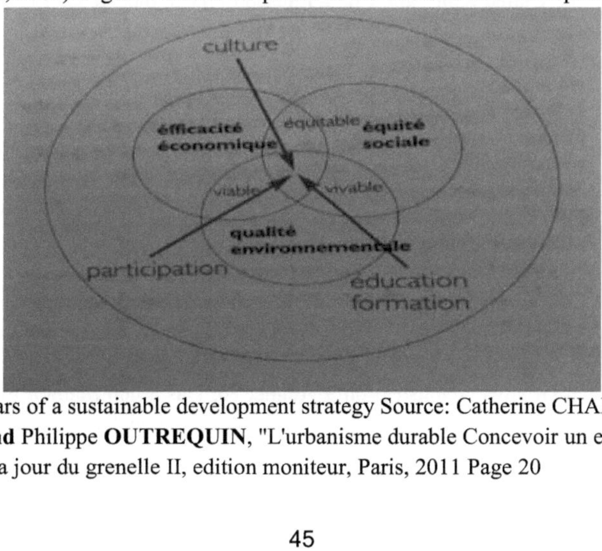


**Figure 6:** Pillars of a sustainable development strategy Source: Catherine CHARLOT-**VALDIEU and** Philippe **OUTREQUIN**, "L'urbanisme durable Concevoir un eco-quartier, 2ieme edition a jour du grenelle II, edition moniteur, Paris, 2011 Page 20

## VII.3. THE COP AND THEIR IMPACT ON CITIES

### VII.3.1. DEFINITION AND OBJECTIVES OF THE COP

The Conference of the Parties (COP) commonly refers to the **annual meeting of** states to set **global climate objectives**. In fact, the COP is a form of organisation specific to certain international conventions. In the field of the environment, there are **3 COPs**, born of 3 conventions signed at the end of the Rio Earth Summit in 1992: the COP on biodiversity, the COP on combating desertification and the COP on climate change, which is the most important of the three. The UNFCCC (United Nations Framework Convention on Climate Change), adopted in 1992, recognises the existence of climate change and **aims to stabilise greenhouse gas** (GHG) **emissions** by proposing **solutions through international cooperation**. The members of the COP on climate change have been meeting annually since 1995. Agreements have been signed within this framework, including the Kyoto Protocol in 1997 and the Paris Agreement in 2015. The Paris Agreement is a global agreement to combat global warming. It aims to contain global warming to well below 2°C above pre-industrial levels by 2100 ( ), and to continue efforts to limit the rise in temperatures to 1.5°C.
All the signatory states of the UNFCCC (197 in 2018) are invited to the annual COP meeting. The official organiser is the UN, but the conference is held in a different country each year. The aim is to find solutions to slow down global warming, monitor the agreements put in place and prevent cities from becoming uninhabitable...

- **Among the decisions taken at the various COPs, the reduction of greenhouse gas emissions, the implementation of the Kyoto agreement and the Paris agreement (Table 5. COPs, host cities and main objectives and agreements. IN APPENDICES).**
- One of the objectives of the COPs following COP11 in Montreal was to reflect on the post-Kyoto period. The binding agreements were due to expire in 2010. It was at COP18 that a second commitment period for the Kyoto Protocol was decided, to run until 2020. It is important to understand that the negotiators' great fear is that they will not succeed in finding a **new** binding **agreement** that will affect all countries, particularly the biggest emitters. COP15 in Copenhagen, for example, was such a disappointment that expectations were so high. In the end, the United States and China refused to make their reduction targets binding, undermining the united fight against climate change. COP15 in Copenhagen did, however, make it possible to affirm that the nations of the world wanted to curb global warming to below +2°C.
- If we add up all the national contributions, it appears that global warming should be +3.5°C by 2100, which **is still a long way from the target set**. The challenge for the post-COP21 period is therefore to get citizens, private and state companies, and local authorities to take part in mitigation actions under the slogan "the climate is everybody's business". Other mechanisms will also be discussed at other COPs, such as the introduction of a carbon price or a green climate fund.
- The 5th IPCC report, published in November 2014, states with certainty that humans are responsible for climate change. The IPCC's "special" report on the consequences of global warming of 1.5°C, published on 6 October 2018, is alarming and calls for radical change. The report by Working Group 1, published on 9 August 2021 as a contribution to the 6th IPCC report (AR6), merely confirms the certainties already acquired by providing a high-level summary of the understanding of the current state of the climate, including how it is changing and the role of human influence. The report by IPCC Working Group 2, published in February 2022 as a contribution to AR6, entitled "Impacts, Adaptation and Vulnerability" is

very alarming. Reports are piling up predicting climate catastrophes for cities and the planet, but everyone is hiding from the truth and emissions are continuing to rise exponentially. A number of measures have been taken, such as the use of the "GREEN EURO", a premium linked to reductions in $CO_2$ emissions and paid to countries that have fought and are still fighting against climate change.

It should be noted that due to the COVID 19 health crisis, there was no COP in 2020.

## VII.4. SUSTAINABLE URBAN DEVELOPMENT OR THE CITY IN THE AGE OF SUSTAINABLE DEVELOPMENT

Sustainable urban development has been defined as *"a process based on collective learning, the ability to resolve conflicts and the desire for a strategic vision, and not on the application of a predefined model. The different systems that make up the city (the economic system, the social system, the physical system and the environmental system) must be considered together, in their coevolution and dynamic interactions (externalities, feed- back, increasing returns, synergies), and not simply the sum of different aspects and objectives".* (**MATHIEU ET AL**, 2011, p. 165).

The sustainable city also obeys well-defined criteria to be able to claim this sustainability. It is therefore important to combine all the components, taking into account their evolution, their dynamics and their impact on the environment.

### VII.4.1. SUSTAINABILITY CRITERIA FOR A CITY

J Reducing the consumption of space ;
J Limiting the unnecessary movement of people and goods;
J Controlling energy consumption and pollution ;
f Improving urban quality ;
f Accessibility for all to the benefits of the city ;
f Promoting health in the city ;
f Promoting heritage ;
f Economic development of the city ;
f Economic and rational management of resources.

These criteria make a city eligible for the rank of sustainable city if they are respected in its management. The COPs take them into account from year to year.

### VII.4.2. THE 1994 AALBORG CHARTER AND ITS PRINCIPLES

The city being the receptacle and the field of action, the inherent discipline is sustainable urban planning. This concept actually came into being with the drafting of the Aalborg Charter in 1994. In May 1994, Aalborg (Denmark) hosted a European conference on **sustainable cities**. Several countries took part, and after various discussions, a charter was unanimously adopted setting out the priorities for the work and a commitment to take part in the Agenda 21 process.

The point to be made when drawing up the Aalborg charter is that it is intended for European cities, whereas the ATHENS charter is intended for any city, whatever its morphology, geographical location or economic impact. The participants at the Aalborg conference signed the following agreement:

*"We, the European towns and cities, signatories to this Charter, undertake by participating in the European Sustainable Towns and Cities Campaign to make every effort to reach a consensus within our communities on a local programme in response to Agenda 21 (Local Agenda 21) by the end of 1996. In this way we will fulfil the task set out in chapter 28 of Agenda 21 adopted by the Rio Summit in June 1992. Through our local initiatives, we will*

*contribute to the implementation of the European Union's Fifth Environmental Action Programme entitled "Towards Sustainability". Local Actions 21 will be defined on the basis of the first part of this charter.*

The plan of this charter is presented in the form of 14 points:

1) The role of European cities ;
2) The concept and principles of sustainability ;
3) Local sustainability plans ;
4) Sustainability, a process of innovation and balance at local level ;
5) Negotiation as a problem-solving method;
6) The urban economy towards sustainability;
7) Social justice for urban sustainability ;
8) Sustainable regional development;
9) Sustainable urban mobility ;
10) Responsibility for global climate change;
11) Preventing the poisoning of ecosystems ;
12) Self-management at local level, a necessary condition for sustainability ;
13) Citizens, protagonists of sustainability, and community participation ;
14) The instruments of sustainability-oriented urban management.

These principles should be respected in all strategic planning.

**Table 6. Comparison between the ATHENS Charter and the Aalborg Charter**

| CHARTERS OF ATHENS(GREECE) 1933 | AALBORG CHARTER (DENMARK) 1994 |
|---|---|
| - The clean slate principle | - Heritage attitude (preserving and enhancing existing assets) |
| • Abstraction of architecture from its surrounding context (history, geography, culture, ecology).<br>• -adoption of the international style | • Keep account of the environment multidimensional and building integration<br>• Architectural diversity |
| - Zoning procedure | - Functional mix procedure and cross-functional policies |
| • Smoother journeys<br>• -Separation of traffic routes | • Reducing forced mobility<br>• Reconquete des voies par les différents modes de transport |
| • Urban planning by experts and technicians<br>• -geometrisation and rationalisation of the city | • Participatory urban planning<br>• Unique responses adapted to the context. |

Source, author, 2023

The elements of comparison and difference between the two charters represent a turning point in the development of urban planning approaches.

On the basis of these principles, town planning and urban development have decided to reconcile these different dimensions within a given area. The notion of sustainable development marks an accentuation of environmental issues in urban production practices. The urban project as an approach has come to intervene at all scales (buildings, urban projects and the city/urban region). For example, at building level, the construction of energy-efficient buildings using environmentally-friendly materials and the implementation of innovative construction techniques, including the use of photovoltaic panels, rainwater harvesting for watering lawns and gardens, natural ventilation, accentuation of light through bay windows, sound and heat insulation using noble materials, double glazing, composting of household waste, selective sorting of household waste, and so on.

On the scale of major urban projects, a move towards renewable energies (solar and wind power), better siting in terms of orientation, waste water treatment and recycling, and good use of flora and fauna through a judicious choice of plants and trees. At the level of the city or urban region, the reorganisation of road infrastructures with multiple use and sharing of lanes, the improvement of ecological public transport, cycle paths, electric scooters, etc. The introduction of blue and green screens, ecological corridors respecting the biodiversity of the ecosystem, rest stops with adapted and ergonomic urban furniture, the establishment of breathing and relaxation areas contributing to the well-being of users.

This growing awareness of the contribution that cities can make to sustainable development led, from the 1990s onwards, to the development of urban projects that attempted to meet sustainability requirements as effectively as possible. The pioneers of this approach are to be found mainly in Northern Europe, where the first "sustainable neighbourhoods" saw the light of day. The creation of sustainable neighbourhoods, renamed "eco-districts", has become a stated objective of many European local authorities concerned with urban sustainability. Success has also spread beyond European borders, with a number of eco-districts in Dubai-City and Qatar taking account of the vagaries of climate and geography.

- **II.5. URBAN PROJECT AND SUSTAINABLE CITY/WHAT RELATIONSHIP?**

Linking urban design and sustainable development means involving urban ecology in research into development through urban design, the quest for well-being and respect for nature. A number of theorists have based their work on the feasibility of the operation and the way to build a flexible model, adaptable to the context and affecting all scales: from the local, to the regional, to territorialisation. The adage "think global-act local" has become the leitmotiv of this thinking.

The aim is to involve all the stakeholders by collaborating with government bodies, private developers, citizens and associations in order to: build **an intentional model** of eco-systemic and contextual sustainability (for the site);

- **set up an iterative diagnostic process** using the intentional model, a kind of framework to be followed, making sure that all the players have been introduced and that the scales and timescale respected in the context of the urban organisation in question ;
- **highlight the sustainability criteria** (the evaluation grid) in order to mark out and frame the process while adapting it to the context and providing coherent actions while taking into account the uncertainties linked to societal evolution, the hazards of life, the climate, natural disasters, etc.

By taking these factors into account, we can ensure that our cities are more sustainable by making the urban design process more collaborative, more transactional and more inclusive of different criteria such as flexibility, temporality, context and spatiality.

Transdisciplinarity and multidisciplinarity are becoming the catalyst for this collaboration with a single objective: to make sustainable cities where the well-being of the inhabitants is the main cause. The paradigm shift is not imposed by imposing radical gestures, but rather follows the speed of the resilience of local communities and the players involved. "*This is one of the many inevitable contradictions of sustainable development: taking the time to devise complex solutions and integrate them into society, while at the same time responding to the sense of urgency generated by the crisis*" (Mertenat, 2012). urbanity offering a mixed and diversified social and natural context, in the form of a compact form of occupation. The current concern is how to reconcile risks with the modelling dynamics of urban eco-systems.

**CONCLUSION**

Sustainable development is an approach that applies to all levels of the territory. It must be (re)defined for each of these scales, with the objectives differing according to the level concerned (from a development project concerning a region to a local housing estate). The issues and objectives are not identical. The methods and tools are different and must therefore be developed for each type of project, for each scale of territory. Eco-districts can be considered as precursors of more sustainable urban development. More than anywhere else, they have given thought to what a sustainable city could be. For the moment, these are relatively isolated projects, characterised by strong political support. Many are new-build projects that extend the existing town, and almost all are part of initiatives in which the local authority's strong financial, organisational and political involvement ( ) has made it possible to remove certain obstacles, in particular those relating to the availability of land and mobility. But these projects also demonstrate that it is possible to produce a more 'sustainable' city, from both a technical and organisational point of view. The challenge now is to understand the extent to which these exceptional approaches can contribute to the emergence of sustainable urban planning or compact cities.

The imperatives of sustainable development are reflected in the process of building these neighbourhoods by concrete technical, programmatic and procedural difficulties. How can economic activity and housing be combined in a project without creating a nuisance? How can we finance the additional costs of energy performance during the construction of the buildings? How can public participation be reconciled with the technicalisation of urban planning and development? To answer these new questions, those involved in producing the city are having to change their professional practices. At the same time, with the addition of new constraints, the context in which the interplay between the players involved in urban planning and development projects takes place is becoming more complex. The introduction of an intentional model that takes account of the various criteria is proving to be one of the solutions for reconciling sustainable development and urban planning.

Various international meetings on climate change are being held all over the world (COP 23 in Egypt, for example), with important recommendations for signatory countries to follow. Sustainable Development comes into play to redirect our efforts towards respect for nature in all its aspects.

## GENERAL CONCLUSION

The first course on the urban project enables the learner to retrace the birth of the concept and the operations which marked its birth. A few reminders and preliminary notions were therefore necessary to introduce the concept in a flexible way (qualities of the urban project) to the vision of traditional planning and the search for conciliatory and non-segregative strategies for all users of space, whether they be residents, decision-makers, developers, financiers or visitors. As the urban project is polysemic, we will first describe the field of action in which it evolves. We have seen that the urban project is an approach that covers all scales: from the building, via the district, to the whole city, and ultimately encompasses the whole territory.

In a bottom-up action, the local level takes the locomotive and drives spatialities across, giving each stop its own set of actions and impacts. Identifying the 'relevant' perimeter of the urban project is not easy, given the scale of the operations. Nevertheless, mastering and managing the "target area" makes the action concrete and the strategy relevant. It calls on interdisciplinarity and transdisciplinarity, is flexible and adaptable in terms of the actions to be taken and how they are to be implemented, adapts to the consequences on the ground, bends to the demands of reality, integrates into processes that have already begun, takes account of general interests by avoiding friction between stakeholders, is democratically open, and creates urban marketing around the project.... In short, this is the bottom-up logic compared with traditional planning.

The urban project is ultimately a flexible planning strategy. It crosses all the scales of an area and is defined according to the relevant perimeter, involving consultation and participation by all, because it concerns everyone.

**The vagueness that surrounds it makes it difficult for theorists to define the urban project as a whole.**

Several approaches have been used, starting with the spatial approach, which looks at the physical transformations of a site, generated by the actions planned and, of course, the resulting changes. Philippe **PANNERAI AND MANGIN** in 1999, Patricia **INGALLINA** in 2002. **ZIMMERMAN** (2000, 2004), **PINSON G.** (2005, 2006, 2009), Sabine **COURCIER** (2002), **HUBERT** et all (2014) and **MARTOUZET** (2018) all advocate an approach based on the stakeholders who have the opportunity to intervene in a project or action and who have the power to make decisions. They all assert that political decisions, the involvement of stakeholders, and their readiness to respond to hazards and resolve obstacles are the keystone of the urban project. The approach incorporates a temporal perspective that retranscribes the urban project in its entirety, and we draw on the work of **ARAB** in 2004, **GROSBELLET** in 2016 and **JOLIVET** in 2018.

the urban project has come to shape the city through major urban projects, calling on the players involved in line with urban policies and regulations in force and their practical implementation. Whether in practice or in teaching, the urban project is a powerful lever for public action and brings together citizens, local elected representatives, investors, bankers and users around its objectives. The actors involved in the action are important because of the political will to undertake work and to stick with it despite the obstacles encountered. According to **PINSON** and **VION** (2000), the urban project is seen as a tool for repositioning the city internationally, a desire to unite urban players in a context of international urban competition. The major urban planning projects, some of which have been discussed in the course content, show that urban policies are above all the decisions of responsible and willing players.

In traditional cities, the small size of the territory and the relative simplicity of the decision-making process mean that the image of the area is perceived as uniform. This is no more self-evident in today's enlarged, fragmented and compartmentalised urban spaces. Methods of representation and decision-making processes must adapt to this new situation. Producing a project on the scale of the urban region will therefore require an 'alliance' between the conurbation and the surrounding peri-urban municipalities. The creation of a development council bringing together all the vital resources of an area is very important to ensure the success of a project. These questions of "governance" are not incidental, because they create the conditions for collective ownership of the project and the coherent projection of its image to the outside world. At other levels, cooperation is essential. Cities are never alone in the world; they are all involved in dynamics of dependence and complementarity with other cities. City-to-city cooperation and the development of their respective assets, within the framework of "city networks", are often the key to mutual attractiveness. Alliances and networking are essential. Nor can cooperation strategies between regions ignore the North-South dimension of their relational system. The circle of relationships is linked to history, to its population, and therefore to its very image. Appropriate political support will be needed to ensure that the proposed forms of cooperation succeed and are adopted by their inhabitants. This is how the urban project could succeed. A tertiary economy, focused on services, information, the quarter, software... has developed with the rise of culture, leisure, urban tourism... These socio-economic transformations of the productive system have an impact on the city by modifying space and usage, scales and territories, pushing the introduction into a new post-industrial urban age and a new urban revolution where the leitmotiv is sustainable development. Sustainable development is an approach that applies to all scales of the territory. It must be (re)defined for each of these scales, with the objectives differing according to the level concerned (from a development project concerning a region to a local housing estate). The issues and objectives are not identical. The methods and tools are different and must therefore be developed for each type of project, for each scale of territory.

Eco-districts can be seen as precursors of more sustainable urban development. More than elsewhere, they have given thought to what a sustainable city could be. For the moment, these are relatively isolated projects, characterised by strong political support. Many are new-build projects that extend the existing town, and almost all of them are part of a process in which the local authority's strong financial, organisational and political involvement has enabled certain obstacles to be overcome, in particular those relating to the availability of land and mobility. But these projects also demonstrate that it is possible to produce a more 'sustainable' city, from both a technical and organisational point of view. The challenge now is to understand the extent to which these exceptional approaches can contribute to the emergence of sustainable urban planning or compact cities.

With the addition of new constraints, the context in which the interplay of players takes place during the implementation of development and urban planning projects is becoming increasingly complex. The introduction of an intentional model that takes account of different criteria is proving to be one of the solutions for reconciling sustainable development and urban planning.

Various international meetings on climate change are being held around the world (COP 23 in Egypt, for example), with important recommendations for signatory countries to follow.

Sustainable development comes into play to redirect our efforts towards respecting nature in all its aspects.
We are asking ourselves the following questions: where is the urban project going? What strategies should be followed to get round the lobbying that would lead to derive? Will the major urban projects (GPV) with teleworking and NICTs one day be obsolete as a means of revitalising the substrate? What will become of these projects built in line with the 'work at home' trend? Does the urban project as an approach still have a future? The answers to all these questions have yet to be invented, and these are avenues for research into PROSPECTIVE, visionary theory.

## BIBLIOGRAPHY IN ALPHABETICAL ORDER

**ARAB, N.** (2004). L'activité de projet dans l'amenagement urbain : processus d'elaboration et modes de pilotage : Les cas de la ligne B du tramway strasbourgeois et d'Odysseum a Montpellier.

**ALVAS J. (1998),** "Bilbao et l'effet Guggenheim", in Le Soir, 30.10.

**ARNAUD, M.** (1996), "Grands Travaux et developpement urbain", *"villes en developpement",* no. 34, December 1996.

**ASCHER, F., (**1995), "*Metapolis, ou l'Avenir des villes*", Paris, Ed. Odile Jacob, 346 pages.

**BOHIGAS,O.** (1985). "Espaces publics et urbanisme de projet : la reconstruction de Barcelone", Amenager l'urbain de Montreal a San Francisco, by Annick Germain and Jean-Claude Marsan 1987, editions du Meridien, pages 65 a 76

**BONNAFOUS A. ET PLASSARD, F. (1974),** " Revue economique ", volume 5 ; numero 2

**BOUCHAREB A.** " Projet Urbain, définitions, acteurs, stratégies ", cours pour les etudiants master 1 academique 2011/2012, departement d'urbanisme, faculte d'Architecture et d'Urbanisme, Universite Salah Boubnider Constantine 3.

**BOURDIN, A**.; **LEFEVRE;M.-P.**&MELE, P. (2006). Les regles du jeu urbain, entre droit et confiance, Descartes & Cie. Les Urbanites

**BOUTINETJ.P.**, (1993a). *"Psychologie des conduites a projet",* Que sais-je no 2770, Paris, PUF, 126 pages.

**BRODHAG,C.**,(2004) *Le developpement durable du concept a l'action* : le SD 21000 Ecole Nationale Superieure des Mines de Saint-Etienne,

**BRUNDTLAND,G. H.**, (1989), *Notre Avenir a Tous,* report of the World Commission on Environment and Development, Les Editions du Fleuve, Paris (French translation of Our Common Future (1987)).

**BRUNEL**, S.,(2004), *Le developpement durable*, Paris, Ed. PUF, 128 pages.

**BUSQUETS,J.**(1990), *"Les projets urbains de niveau intermediaire : l'exemple de Barcelone",* trames n°1, volume 3.

**BUSQUETS**, J. (1991). *"La planification-cadre et les projets-actions",* Les Annales de la recherche urbaine no 51 juillet 1991, pages 123 a 130.

**CABET, E.,** 1840*, Voyage en Icarie*, Paris.

**CAGNARDI, Cesare MACCHI CASSIA. Vittorio SPIGAI,**(2015) " Planification et projet urbain en Italie in CRU (centre de documentation de l'urbanisme) Paris,

**CAMPOS,G. VENUTI, COSTA P.L., PIAZZA, L., ODOARDO REALI.ED MARSILIO LUIGO** .(1985), conference. "Florence for quality urban planning. Preliminary draft of a master plan" Venice, P.236

**CHADOIN, O. et AL** (2015) EDITORIAL in Espaces et Societes n°163 page 7. Published online on Cairn.info on 06/11/2015https://doi.org/10.3917/esp.163.0007

**CHALINE,CL.** (1999), "la regeneration urbaine", Paris.PUF. Que sais-je n°3496, 127 pages.

**CHARLOT-VALDIEU,C.** and **OUTREQUIN,P.**,(2006). *Developpement durable et renouvellement urbain " des outils operationnels pour ameliorer la qualite de vie dans nos quartiers ",* edition d'Harmattan, France, P: 15.

**COURCIER,**S. (2002) "de revaluation de l'effet structurant d'un projet urbain a l'analyse des congruences entre stratégies d'acteurs : le reamenagement du vieux port de Montreal", doctoral thesis, University of Montreal, Canada.

**DEMAZIERES C. ET RODRIGUEZ, A. (1999),** "Le grand projet emblematique, reponse

Des metropoles a la globalisation. Le cas de Bilbao", *Infrastructures et formes urbaines*. Paris L'Harmattan, Espace et societe no 95, pages 147 a 160.

**DEVILLIERS,C.**, (1988). "*De la logique de secteur au projet urbain*", Interview by P. Genestier and M. Roncayolo, pages 244 to 259.

**DEVILLIERS, C., GIRALDEAU F.**( 1990), "Notes sur le projet urbain :enjeux et methodes", Trames n°1 volume 3, pages 6 to 12.

**DEVILLERS, C.**, (1990). "*Y-a-t-il un projet dans la ville*?", Plan Construction.
Séminaire de prospective sur l'habitat et les villes, December 1990, 10 pages.

**DEVILLIERSC.**, (1994), " *Le projet urbain* ", Conferences Paris d'Architectes 1994 au Pavillon de l'Arsenal, les minis PA no2, editions du Pavillon de l'Arsenal, 1996, pages 7 to 47.

**GIRALDEAU, F.** (1990). "Notes sur le projet urbain : enjeux et methodes", Trames no 1 volume 3, pages 6 to 12.

**GENESTIER,P.**(1993), "*Que vaut la notion du projet urbain?* in l'architecture d'aujourd'hui, n°288.

**GENESTIER**, P. (1996). Towards a new urbanism. Faire la ville, comment? Paris, La documentation fran^aise.

**GENESTIER, P.**(2001). Des projets en parole et en images. La rhetorique du projet face a la crise du vouloir-politique. Espaces et Societes, n° 105-106, pp. 101-126

**GROSBELLET, J.** (2016). La ville acceleree et ses acteurs : Anatomie du projet urbain des Bassins a flot a Bordeaux. Universite Paris-ouest, Naterre La defense.

**HAUMONT, A.**, (1993 ), La mobilite intra-urbaine : Retrospective et prospective **.**inLes Annales de la Recherche UrbaineAnnee 1993 59-60pp. 109-118

**HARMELLE, C.,** (1982), Revue Recherche n° 47 /-48 /Les piques de l'aigle. librairiephilippearnaiz, Isle sur la sorgue, France

**HAYOT,A.** (1997). "En finir avec le fonctionnalisme : les sciences sociales, l'architecture et la ville", Alain Hayot and Andre Sauvage, eds. (2000). Le projet urbain, enjeux experimentations et professions. Actes du colloque "les sciences humaines et sociales face au projet urbain" organisée par l'INAMA et SHS-TEST a Marseille en janvier et fevrier 1997. Paris : editions de la Villette, pages 11 a 24

**HECHAM ZEHIOUA, B.** , (2010), "Impacts des projets inscrits a Constantine et evaluation de son image de marque, pour un projet urbain a effet structurant", PhD thesis Es Sciences, defended in May 2010, Mentouri Constantine 1 University.

**HUBERT, M.** et **al** (s/d ) (2014). Les Grands projets urbains : Territoires, acteurs et strategies. Presses universitaires de Montreal

**HUET, B.**(1986), "l' architecture contre la ville", *Architecture mouvement continuite (AMC) n°14* decembre 1986, pages 10 to 13.

**INGALLINA, P.** (2001) " Le projet urbain " ED presses universitaires de France, collection : Que sais-je ? N°3585, Paris, 127 pages

**JOLIVET, D.** (2018). Maturite des projets d'urbanisme: vers une structure temporelle du projet. ln Les grands projets urbains : Territoires, acteurs et strategies, (s/d) D. Martouzet. pp.159- 193

**LAPERRIERE, H.** (1998). "Lecture culturelle d'un paysage urbain : elements de methode tires du cas montrealais", Lieux culturels et contextes de villes sous la direction de J.P. Augustin et D. Latouche, Maison des sciences de l'Homme d'Aquitaine, pages 113 to 135.

**LE PETIT LAROUSSE 2019.**
**MANGIN,D. PANNERAI,P.** (1999), Projet urbain " ED. Parentheses, 186 pages
**MARSAN, J-CL** (1994) Montreal in evolution. Laval : Quebec, Editions du Meridien, 515 pages.
**MARTOUZET, D.** (2018). Le projet fait les acteurs : Urbanisme, complexite, incertitude. s/d D. Martouzet. Presses universitaires Francois-Rabelais.
**MASBOUNGI, A. (2016). Le plaisir de l'urbanisme. Editions parentheses. 176p.**
**MASBOUNGI A.** (2002), "Projets urbains en France", Moniteur, Paris
**MATHIEU,N.**, and **GUERMOND,Y.**, (2011). *La ville durable, du politique au scientifique*, edition Quae, France,. P : 165.
**MERLIN,P.** and **CHOAYF**. (1996), "Dictionnaire de l'urbanisme et de l'amenagement", Paris, PUF, 863 pages.
**MOLLURA**, **G.**(2011)" Les programmes europeens et les villes: politiques integrees de requalification urbaine et 'gouvernance'", *Rives mediterraneennes*, 39 | 2011, 113-123.
**MONNIER,E.** (1992). *Evaluation de l'action despouvoirspublics*. Paris : Economica 245 p
**OFFNER, J-M** (1993). "Les effets structurants du transport : mythe politique, mystification scientifique". L'espace geographique, no3 1993, pages 233 a 242
**OFFNER, J-M** (1995). "L'invention politique et sociale de la technique", Ville espace et valeurs under the direction of J.L. Gourdon et al. Paris l'Harmattan, pages 521 to 534.
**PINSON,D.** (1997) ," *Projet de ville et projets de vie* ",Hayot, Alain et Andre Sauvage, dir.(2000) du " projet urbain, enjeux, experimentations et professions ".acte du colloque " les sciences humaines et sociales face au projet urbain " organisé par l'INAMA et SHD-TEST a Marseille en janvier et fevrier 1997. Paris. Editions de la Villette, pages 78 to 95.
**PINSON, G. AND VION, A.** (2000). L'internationalisation des villes comme objet d'expertise, Pole Sud, n°13, pp. 85-102.
**PINSON, G**. (2002). Projects and powers in European cities. A comparison of Marseille, Venice, Nantes and Turin. These de doctorat, Universite de Rennes I
**PINSON, G.** (2005a). Le projet urbain comme instrument d'action publique, in P. Lascoumes et al, Gouverner par les instruments, Presses de Sciences Po, pp.199-233.
**PINSON, G.** (2005b). L'ideologie des projets urbains. L'analyse des politiques urbaines entre precedent anglo-saxon et 'detour' italien. In Sciences de la societe, n°65. "Les ideologies emergentes des politiques territoriales. pp. 28-51
**PINSON, G.** (2009). Gouverner la ville par projet : urbanisme et gouvernance des villes europeennes, Presses de Science Po, Paris, 418 p.
**PINSON, G**. (2006). Projets de ville et gouvernance urbaine : pluralisation des espaces politiques et recomposition d'une capacité d'action collective dans les villes europeennes Sciences Po. "Revue francaise de science politique. Vol. 56, pp. 619-651 **PINSON, G.** (2009). Gouverner la ville par projet : urbanisme et gouvernance des villes europeennes, Presses de Science Po, Paris, 418 p.
**POUYANNE** G., 2004, *urban form and daily mobility,* thesis for doctorate in economics, Universite Montesquieu-Bordeaux IV, December, p. 12.
**REMY,J.** (1998), *"Le projet urbain sens et signification",* Toussaint Jean-Yves and Monique Zimmermann, dir. De " projet urbain, menager les gens, amenager la ville ",Pierre Mardaga editeur,p.5 a 8.
**REVIEW INTERNATIONAL OF SCIENCES OF MANAGEMENT**
ISSN : 2665-7473 Volume 4 : Numero 1 Revue ISG www.revue-isg.com Page 85

**SOKOLOFF,B.** (1999). Barcelona or how to remake a city. Les Presses de l'Universite de Montreal, 208 pages.

**STEINEBACH,M.** (1993). "*Reflexions sur le projet urbain et le role de la forme urbaine, de l'espace dans la relation Mutation Economique et Urbanisme*", Lacaze, (dir. de) Mutations economiques et urbanisation, Paris: La Documentation fran^aise, p357 a 369.

**THOMAS F.** (1998). "*Vers une nouvelle culture de l'amenagement des villes*", J. Y. Toussaint and M. Zimmermann (eds.). Projet urbain, menager les gens, amenager la ville, Pierre Mardaga editeur, p15 a 34

**THOMAS F.,** (1998). Cited by S. Courcier in "Canadian Journal of Urban Research Vol. 14, No. 1, Supplement: Canadian Planning and Policy / Amenagement et politique au Canada (2005), pp. 57-80 (24 pages), Published By: Institute of Urban Studies, University of Winnipeg.

**TOUSSAINT J.-Y. & ZIMMERMANN M.** (dir.), *User, Observer, programmer et fabriquer I'espace public,* Collection des sciences appliques de 1'INSA de Lyon, 2001

**VERHAGE R. et LEROY M.** , " Developpement urbain durable : Comment apprendre des expériences d'ecoquartiers ? ", *Geocarrefour* [En ligne], vol. 89/4 | 2014, online since , consulted on 03 January 2023. URL : http://journals.openedition.org/geocarrefour/9558 ; DOI : https://doi.org/10.4000/geocarrefour.9558

**WILLIAMSON, J.** (1995). Community design management, 210 pages

**WCED** (World Commission on Environment and Development), 1987, *Our Common Future*, Oxford: Oxford University Press, 416 p

**WRENN**, D., M. (1983). Urban waterfront development. Washington, the Urban Land Institute, 219 pages **https://ec.europa.euhttps://www.ibicity.fr)**

https://www.yumpu.com/fr/document/view/34692540/planification-et-projet-urbain-en-italie-documentation-centre

https://www.compteco2.com/article/historique-cop-conference-des-parties

## SECTIONS B. TUTORIALS/PROJECTS HIGHLIGHTED

We divided the tutorials into two sub-sections: the global sub-section, in which a few experiences from around the world were studied and presented by the students, and a local sub-section: experiences in Algeria.

Based on the principle of the general to the local, we have drawn up these two subsections.

### B.1. FIRST SUBSECTION//

The choice of topics is left up to the learners, while providing a professional framework for the themes and example chosen.

In this way, major urban planning projects around the world using the urban project as an approach to action were targeted.

The diversity of the projects enriches the culture of urban design.

We just need to check the criteria that differentiate an "urban project" approach from a traditional directed planning approach.

The operation as such is subject to a very precise work plan: taking account of the geographical situation, the operation itself, the players involved, public consultation, public/private partnerships, the results of the implementation, the post-realisation impact and the addition of a comparison with our own country: can these operations be transposed or replicated in Algeria?

### B.1.1. TITLES OF GLOBAL PROJECTS CHOSEN BY STUDENTS

* Some project titles submitted by students and not included in this collection.

B.1.1.1 Bologna's historic centre: a pioneering urban project

B.1.1..2. Development of the Clyde Valley in Glasgow (Great Britain)

B.1.1.3.. THE CITY OF Malmo ENsuede

B.1.1.4.. THE REDEVELOPMENT OF THE OLD PORT Montreal Canada

B.1.1.5.. Conversion of Saint Denis-France station

B.1.1.6...Eco-neighbourhoods in German cities

B.1.1.7 Redevelopment of the Confluence sector of Lyon-France

B.1.1.8..barceloneluenc@22 in Spain

B.1.1.9... The Bilbao Urban Project (Spain)

B.1.1.10. The Eco-neighbourhood concept in the United Arab Emirates

B.1.1.12.. La rose de Cherbourg , La defense-paris district France

B.1.1.13..Le grand Mirail in Toulouse France

B.1.1.14. High line new YorkUnited States

B.1.1.15.. The Euralille project in France.

Other projects were discussed, but of lesser **importance.**

### 8.2. SUB-SECTION 2/// URBAN PROJECTS IN ALGERIA.

**The TD statement.**

""Having studiedthe projects urban projects cities the world,
-having exposing and discussed the content of each urban urban project,

* after expressing your opinions on a given project in a free and democratic debate;

we are launching the content of the next TD. **"MY CITY IN ACTION".**

* This TD looks at projects in our country: "from internationalisation to nationalisation and localisation".

The content of CE TD is typically planted **in the students' home towns.**

* During the winter and spring holidays, students return home with authorisation from the department to visit, consult and talk to the local players involved in the projects.

* The aim is: to identify **a city project**, a **transformation in one of the city's neighbourhoods**, an **operation to improve the city's urban living environment, a large-scale operation that affects the entire city IN SHORT, AN ACTION IN THE CITY OF RESIDENCE OR A NEARBY TOWN....**

* **Contact the players involved (president of the local authority, head of the town planning department, sectoral managers with a link to the project or action), ask these local players questions in the form of an interview about the project, the choice of project and land, its usefulness, its objectives, its social, economic and town planning impacts... on the town and its future; * Contact the population concerned by this action and ask whether there has been any consultation or awareness-raising... whether this population agrees with the action or not.Contact the population concerned by this action and ask whether there has been consultation or awareness-raising... whether this population agrees with the action or not, the whys and wherefores... Y' is citizen participation?**

* **Of course, you need to carry out an urban analysis of the site where the action is taking place, highlighting the changes and impacts, and carrying out a SWOT analysis,**

* **Once the work is done, draw up a summary and assessment of this action, argue whether this operation complies with the criteria of the urban project and put forward your own point of view in the form of a development proposal with the slogan "If I were the developer, I would have done....) with supporting arguments.**

### 8.2.1. TITLES OF SOME PROJECTS

8.2.1.1. The Constantine tramway and its extensions

8.2.1.2. The BARDO -PARK project in Constantine

8.2.1.3. Conversion of the fishing port of SKIKDA

8.2.1.4. Conversion of the EDIMCO district of Bedjaia

8.2.1.5. Uganda Refinry Project (I'etudiant est du pays de 1'OUGANDA)

8.2.1.6. The development of the industrial zone and its impact on the town of Bouira

8.2.1.7. The permanent safeguard plan for old Mila (the **PPSMVSS)**

8.2.1.8. The industrial zone of the town of BORD-BOUARRERIDJ and its radius of influence B.2.1.9. The RIDE tourist circuit project in southern Algeria

8.2.1.10. The petrochemicals port in Skikda

8.2.1.11. MILA bus station

8.2.1.12. The sports complex in Tebessa.

8.2.1.13. SNOBER LAND, a dynamic theme park for the town of EL Khroub B.2.14. L.UC3, an urban project with a structuring effect

8.2.1.14. Constantine, Capital of Arab Culture 2015, an eventful urban project.

There are other projects, but of lesser importance or concerning the same town.

APPENDICES

**Table 5. The different COPs, city, organising countries and main objectives.**

| Year | COP | CITY | COUNTRY | CONTENTS |
|---|---|---|---|---|
| 1995 | COP 1 | Berlin | Germany | -reduction in greenhouse gases |
| 1996 | COP2 | Geneva | Switzerland | -climate change: a threat to the survival of the planet. |
| 1997 | COP3 | Kyoto | Japan | -signature of binding protocol: reduction of 5.2% of greenhouse gases for countries signatories |
| 1998 | COP4 | Buenos Aires | Argentina | -negotiations between countries |
| 1999 | COP5 | Bonn | Germany | negotiations between countries |
| 2000 | COP6 | La Hague | Netherlands | negotiations between countries |
| 2001 | COP6 | Bonn | Germany | negotiations between countries |
| 2001 | COP7 | Marrakech | Morocco | negotiations between countries |
| 2002 | COP8 | New Delhi | India | negotiations between countries |
| 2003 | COP9 | Milan | Italy | The main **objectives** will be to create a |
| | | | | global carbon tax system, the end of coal dependency, the phasing out of fossil fuels and the end of investment in the grey or brown economy negotiations between the country. **clarification of № work of this protocol ;** creationof a global system of tax on carbon |
| 2004 | COP10 | Buenos Aires | Argentina | negotiations between countryaccounting |

| | | | | methodology |
|---|---|---|---|---|
| 2005 | COP11 | Montreal | Canada | -ratification of the of KYOTO |
| 2006 | COP12 | Nairobi | Kenya | Two countries issuers (China and USA) have not signed. |
| 2007 | COP13 | Bali | Indonesia | very convincing negotiations (equity mechanisms) |
| 2008 | COP14 | Poznan | Poland | further of negotiations (implementation of the |
| | | | | observation and governance system |
| 2009 | COP15 | Copenhagen | Denmark | implementation of the negotiated protocolProtect 30 of the planet by 2030 |
| 2010 | COP16 | Cancun | Mexico | end of constraints for for signatories |
| 2011 | COP17 | Durban | South Africa | Canada withdraws from the KYOTO Protocol |
| 2012 | COP18 | Warsaw | Poland | |
| 2013 | COP19 | Doha | Qatar | |
| 2014 | COP20 | Lima | Peru | |
| 2015 | COP21 | Paris | France | Its main elements **are** as follows: long-term objective - governments **have** agreed to contain the rise in average global temperature to well below 2°C by compared to current levels. levels and to continue action taken to limit temperature rise limited to 1.5°CWarming limited to 1.5° The proposed of limitthe |

| | | | | |
|---|---|---|---|---|
| | | | | a global warming target of 1.5° was called for |
| | | | | by small island states, threatened of submersion. ... **all States have undertaken to publish their The United Nations has adopted its own greenhouse gas (GHG) emission reduction targets. They are now available on the United Nations website.** 2) Use of fossil fuels **and** carbon tax. ... 3) Green fund. ... 4) Human rights. ... 5) Involvement of civil society players. |
| 2016 | COP22 | Marrakech | Morocco | defining an action plan by report on the signing of the Paris Agreement at COP 21. |
| 2017 | COP23 | Bonn | Germany | to increase their commitments to reduce from greenhouse gas emissions. It's a waste of time |
| 2018 | COP24 | Katowice | Poland | to enhance their commitments to reduce of greenhouse gas emissions. It's a waste of time |
| 2019 | COP25 | Madrid | Spain | **The aim of COP25** in Madrid was to reach agreement on a final, fairly complex |

|  |  |  |  | rule in the agreement, concerning the mechanisms for trading greenhouse gas emission reductions. |
|---|---|---|---|---|
| 2021 | COP26 | Glasgow | Scotland | the EUa encourages other parties to step up their commitments and actions to reduce the emissions and increase the adaptation efforts to ensure that the objectives of the Paris Agreement remain accessible |
| 2022 | COP 27 | Charm el-Sheikh | Egypt | Responding to climate emergenciesSouthern states hard hit by by |
|  |  |  |  | global warming will require more at countries historically responsible for global warming. They request the respect of financial commitments already taken and more The Paris Agreement provided for everything, i.e. an evaluation mechanism, called the "Assessment". Assessment World Conference, which will be held for the first time at |

| | | | | |
|---|---|---|---|---|
| | | | | COP28 in 2023. The Paris Agreement provided for everything, i.e. an assessment mechanism called the Assessment World Cup, which will be held for the first time at COP28 in 2023. |

Source : https://www.compteco2.com/article/historique-cop-conference-des-parties

Updated by 1'auteur, 2023.

MIX
Papier aus verantwortungsvollen Quellen
Paper from responsible sources
FSC® C105338

Printed by Books on Demand GmbH, Norderstedt / Germany